Contemporary Military Innovation

This book explores contemporary military innovation, with a particular focus on the balance between *anticipation* and *adaptation*.

The volume examines contemporary military thought and the doctrine that evolved around the thesis of a transformation in the character of war. Known as the Information Technology Revolution in Military Affairs (IT-RMA), this innovation served as an intellectual foundation for the U.S. defence transformation from the 1990s onwards. Since the mid-1990s professional ideas generated within the American defence milieu have been further disseminated to military communities across the globe, with huge impact on the conduct of warfare.

With chapters written by leading scholars in this field, this work sheds light on RMAs in general and the IT-RMA in the U.S. in particular. The authors analyse how military practice and doctrines were developed on the basis of the IT-RMA ideas, how they were disseminated, and the implications of them in several countries and conflicts around the world.

This book will be of much interest to students of strategic studies, defence studies, war and technology, and security studies in general.

Dmitry (Dima) Adamsky is Assistant Professor at the Lauder School of Government, Diplomacy and Strategy at the IDC Herzliya, and Affiliate at the National Security Studies Program at Harvard University.

Kjell Inge Bjerga is Assistant Professor and head of Civil–Military Relations at the Norwegian Institute for Defence Studies in Oslo.

Cass Military Studies

Intelligence Activities in Ancient Rome
Trust in the Gods, But Verify
Rose Mary Sheldon

Clausewitz and African War
Politics and Strategy in Liberia and Somalia
Isabelle Duyvesteyn

Strategy and Politics in the Middle East, 1954–60
Defending the Northern Tier
Michael Cohen

The Cuban Intervention in Angola, 1965–1991
From Che Guevara to Cuito Cuanavale
Edward George

Military Leadership in the British Civil Wars, 1642–1651
'The Genius of this Age'
Stanley Carpenter

Israel's Reprisal Policy, 1953–1956
The Dynamics of Military Retaliation
Ze'ev Drory

Bosnia and Herzegovina in the Second World War
Enver Redzic

Leaders in War
West Point Remembers the 1991 Gulf War
Edited by Frederick Kagan and Christian Kubik

Khedive Ismail's Army
John Dunn

Yugoslav Military Industry 1918–1991
Amadeo Watkins

Corporal Hitler and the Great War 1914–1918
The List Regiment
John Williams

Rostóv in the Russian Civil War, 1917–1920
The Key to Victory
Brian Murphy

The Tet Effect, Intelligence and the Public Perception of War
Jake Blood

The US Military Profession into the 21st Century
War, Peace and Politics
Edited by Sam C. Sarkesian and Robert E. Connor, Jr.

Civil–Military Relations in Europe
Learning from Crisis and Institutional Change
Edited by Hans Born, Marina Caparini, Karl Haltiner and Jürgen Kuhlmann

Strategic Culture and Ways of War
Lawrence Sondhaus

Military Unionism in the Post Cold War Era
A Future Reality?
Edited by Richard Bartle and Lindy Heinecken

Warriors and Politicians
U.S. Civil–Military Relations under Stress
Charles A. Stevenson

Military Honour and the Conduct of War
From Ancient Greece to Iraq
Paul Robinson

Military Industry and Regional Defense Policy
India, Iraq and Israel
Timothy D. Hoyt

Managing Defence in a Democracy
Edited by Laura R. Cleary and Teri McConville

Gender and the Military
Women in the Armed Forces of Western Democracies
Helena Carreiras

Social Sciences and the Military
An Interdisciplinary Overview
Edited by Giuseppe Caforio

Cultural Diversity in the Armed Forces
An International Comparison
Edited by Joseph Soeters and Jan van der Meulen

Railways and the Russo-Japanese War
Transporting War
Felix Patrikeeff and Harold Shukman

War and Media Operations
The US Military and the Press from Vietnam to Iraq
Thomas Rid

Ancient China on Postmodern War
Enduring Ideas from the Chinese Strategic Tradition
Thomas Kane

Special Forces, Terrorism and Strategy
Warfare By Other Means
Alasdair Finlan

Imperial Defence, 1856–1956
The Old World Order
Greg Kennedy

Civil–Military Cooperation in Post-Conflict Operations
Emerging Theory and Practice
Christopher Ankersen

Military Advising and Assistance
From Mercenaries to Privatization, 1815–2007
Donald Stoker

Private Military and Security Companies
Ethics, Policies and Civil–Military Relations
Edited by Andrew Alexandra, Deane-Peter Baker and Marina Caparini

Military Cooperation in Multinational Peace Operations
Managing Cultural Diversity and Crisis Response
Edited by Joseph Soeters and Philippe Manigart

The Military and Domestic Politics
A Concordance Theory of Civil–Military Relations
Rebecca L. Schiff

Conscription in the Napoleonic Era
A Revolution in Military Affairs?
Edited by Donald Stoker, Frederick C. Schneid and Harold D. Blanton

Modernity, the Media and the Military
The Creation of National Mythologies on the Western Front 1914–1918
John F. Williams

American Soldiers in Iraq
McSoldiers or Innovative Professionals?
Morten Ender

Complex Peace Operations and Civil Military Relations
Winning the Peace
Robert Egnell

Strategy and the American War of Independence
A Global Approach
Edited by Donald Stoker, Kenneth J. Hagan and Michael T. McMaster

Managing Military Organisations
Theory and Practice
Edited by Joseph Soeters, Paul C. van Fenema and Robert Beeres

Modern War and the Utility of Force
Challenges, Methods and Strategy
Edited by Jan Angstrom and Isabelle Duyvesteyn

Democratic Citizenship and War
Edited by Yoav Peled, Noah Lewin-Epstein and Guy Mundlak

Military Integration after Civil Wars
Multiethnic Armies, Identity and Post-Conflict Reconstruction
Florence Gaub

Military Ethics and Virtues
An Interdisciplinary Approach for the 21st Century
Peter Olsthoorn

The Counter-Insurgency Myth
The British Experience of Irregular Warfare
Andrew Mumford

Europe, Strategy and Armed Forces
Towards Military Convergence
Sven Biscop and Jo Coelmont

Managing Diversity in the Military
The Value of Inclusion in a Culture of Uniformity
Edited by Daniel P. McDonald and Kizzy M. Parks

The US Military
A Basic Introduction
Judith Hicks Stiehm

Democratic Civil–Military Relations
Soldiering in 21st-Century Europe
Edited by Sabine Mannitz

Contemporary Military Innovation
Between Anticipation and Adaptation
Edited by Dmitry (Dima) Adamsky and Kjell Inge Bjerga

Contemporary Military Innovation

Between anticipation and adaptation

Edited by Dmitry (Dima) Adamsky and Kjell Inge Bjerga

Routledge
Taylor & Francis Group
LONDON AND NEW YORK

First published 2012
by Routledge
2 Park Square, Milton Park, Abingdon, Oxon OX14 4RN

Simultaneously published in the USA and Canada
by Routledge
711 Third Avenue, New York, NY 10017

Routledge is an imprint of the Taylor & Francis Group, an informa business

British Library Cataloguing in Publication Data
A catalogue record for this book is available from the British Library

Library of Congress Cataloging in Publication Data
Contemporary military innovation: between anticipation and adaptation/edited by Dima Adamsky and Kjell Inge Bjerga.
p. cm. – (CASS military studies)
Includes bibliographical references and index.
1. Military art and science. 2. United States–Armed Forces–Effect of technology on. 3. Military doctrine–United States. I. Adamsky, Dima. II. Bjerga, Kjell Inge.
U104.C67 2012
355.6'867–dc23 2011052007

ISBN: 978-0-415-52336-3 (hbk)
ISBN: 978-0-203-11254-0 (ebk)

Typeset in Baskerville
by Wearset Ltd, Boldon, Tyne and Wear

Printed and bound in the United States of America by Publishers Graphics, LLC on sustainably sourced paper.

Contents

Figures and table

Figures

Table

Contributors

Dmitry (Dima) Adamsky is Assistant Professor at the Lauder School of Government, Diplomacy and Strategy at the IDC Herzliya. His research interests include international security, modern military thought, cultural approach to International Relations, nuclear weapons and strategy. He has published on these topics in academic journals, in edited volumes and encyclopaedias. His first book *Operation Kavkaz* (Hebrew) earned the prize for the best academic work on Israeli security in 2006. His second book, *The Culture of Military Innovation* was published by Stanford University Press in 2010.

Kjell Inge Bjerga is Assistant Professor at the Norwegian Institute for Defence Studies in Oslo, where he has headed the Departments of Norwegian Security Policy and Civil–Military Relations. He holds a Cand. Philol. Degree from the University of Oslo, and is specialized in the evolution of military organizations and politico–military relations in small states. He has published in academic journals, edited volumes and encyclopaedias, as well as monographs, including *Unity as a Weapon* (Norwegian), Bergen: Eide, 2002. He was editor and co-author of the *Norwegian Joint Operational Doctrine* (2007) and has lectured for 12 years at the Norwegian Defence University College.

Itai Brun is the head of the Research Division in the Israel Defence Forces' Directorate of Military Intelligence. In his previous positions he headed the DADO Centre for Interdisciplinary Military Studies, the Analysis Department in the Israeli Air Force Intelligence and has been deputy head of the Research Division in the Directorate of Military Intelligence. He earned his LLB from Haifa University (cum laude) and he also has a Master's Degree in Political Science (Diplomacy and Security Studies) from Tel Aviv University (cum laude). He has published academic articles and contributed book chapters on military strategy, intelligence and airpower.

Antulio J. Echevarria II is Professor and the Director of Research at the U.S. Army War College. He holds a Ph.D. in history from Princeton

University, and is a graduate of the U.S. Army War College. He has had a distinguished career in the U.S. Army, and has written extensively on military history and military theory. His most recent books include *Clausewitz and Contemporary War*, Oxford, 2007; *Imagining Future War: The West's Technological Revolution and Visions of Wars to Come, 1880–1914*, Praeger (2007); *After Clausewitz: German Military Thinkers before the Great War*, Kansas UP, 2001.

Theo Farrell is Professor of War in the Modern World in the Department of War Studies at King's College, London. His recent books include: *The Norms* of War, Lynne Rienner, 2005; as co-author, *International Law and International Relations*, Cambridge UP, 2007; as co-editor, *A Transformation Gap? American Innovations and European Military Change*, Stanford UP, 2010; and as editor, *Security Studies*, 5 volumes, Routledge, 2010.

Azar Gat is Ezer Weitzman Professor for National Security in the Department of Political Science at Tel Aviv University, which he chaired in 1999–2003. His publications include: *The Origins of Military Thought: The Enlightenment to Clausewitz*, Oxford UP, 1989; *The Development of Military Though: The Nineteenth Century*, Oxford UP, 1992; *Fascists and Liberal Visions of War: Fuller, Liddell Hart, Douhet and Other Modernists*, Oxford UP, 1998; and *British Armour Theory and the Rise of the Panzer Arm: Revising the Revisionists*, Macmillan, 2000. His *War in Human Civilization* was published by Oxford in 2006 and was named one of the best books of the year by the *Times Literary Supplement.*

Torunn Laugen Haaland is Associate Professor at the Norwegian Institute for Defence Studies and Dean at the Norwegian Defence University College. She holds a Ph.D. from the University of Oslo and is specialized in military reforms, developments in role perceptions within the Armed Forces after the Cold War and learning in military organizations. She has published academic monographs, articles and contributed to edited volumes.

Rolf Hobson is Professor at the Norwegian Institute for Defence Studies. He has published numerous academic articles and books. He is specialized in European history, in particular German military history, and is currently working on post-war German political history. He is the author of *Imperialism at Sea: Naval Strategic Thought, the Ideology of Sea Power and the Tirpitz Plan*, Brill, 2004.

Harald Høiback is a Lieutenant Colonel, and lectures at the Norwegian Defence University College. He holds a Masters Degree in Philosophy from the University of Oslo, a Master's Degree in History from the University of Glasgow (M.Phil.), and a Ph.D. in Philosophy from the University of Oslo (2010). In his Ph.D. he discussed the epistemological justification of military doctrine. He has published academic monographs and articles and contributed to edited volumes.

Jacqueline Newmyer Deal is President of the Long Term Strategy Group, a defense research firm in Cambridge, Massachusetts. Recent articles include 'Oil, Arms, and Influence: The Indirect Strategy Behind Chinese Military Modernization', *Orbis*, Spring 2009, and 'When the CCP Loses the Mandate of Heaven', *World Politics Review*, November 2009. Recent testimony includes her appearance at the '30 April 2009 U.S. China Economic and Security Review Commission Hearing on China's Propaganda and Influence Operations, Its Intelligence Activities that Target the United States, and the Resulting Impact on U.S. National Security'.

Stephen Peter Rosen is the Beton Michael Kaneb Professor of National Security and Military Affairs at Harvard University. He was the civilian assistant to the director, Net Assessment in the Office of the U.S. Secretary of Defense, the Director of Political–Military Affairs on the staff of the National Security Council, and a professor in the Strategic Department at the U.S. Naval War College. He is the author of *Winning the Next War: Innovation and the Modern Military*, Cornell UP, 1992; *Societies and Military Power: India and its Armies*, Cornell UP, 1996; and of *War and Human Nature*, Princeton UP, 2005.

James A. Russell is Assistant Professor in the Department of National Security Affairs at the U.S. Naval Postgraduate School in Monterey, CA. He holds a Ph.D. in War Studies from the University of London. His book *Innovation, Transformation and War: Counterinsurgency Operations in Anbar and Ninewa, 2005–2007*, was published by Stanford UP in 2011.

Carmit Valensi is a Ph.D. candidate in the School of Political Science at Tel Aviv University. Her dissertation explores Hamas, Hezbollah and Al-Qaida as 'Hybrid Actors'. She specializes in contemporary Middle East, strategic studies, terrorism and counterinsurgency. She has published in academic and professional periodicals. She has been a visiting research fellow at the Fox Fellow Program for International and Area Studies of 2010–2011 at Yale University, senior research fellow at the DADO Centre for Interdisciplinary Military Studies, and served in the Intelligence Corps of the IDF.

1 Introduction

Dmitry (Dima) Adamsky and Kjell Inge Bjerga

Contemporary military thought and innovation evolved around the thesis about the transformation in the character of war. Known in professional circles as an Information Technology Revolution in Military Affairs (IT-RMA), it served as an intellectual foundation for the U.S. defense transformation. Moreover, the IT-RMA became an umbrella term for a whole raft of military visions, doctrines and concepts, such as 'effect-based operations' and 'network-centric warfare'. Since the mid-1990s professional ideas generated within the American defense milieu were further disseminated to military communities across the globe. RMA turned into an integral part of the professional military lexicon worldwide. In many ways an intellectual history of the IT-RMA encapsulates the development of contemporary military thought.

Several unique features of the IT-RMA make it relevant and important for scholars and students of strategic studies, and for decision-makers dealing with strategy and doctrine development as well as defense planning and procurements. The intellectual history of the IT-RMA is quite puzzling. The Soviets, the Americans and the Israelis, three pioneers in the field, approached this innovation in quite different ways. Whereas in the American and Israeli cases the cultivation of the technological seeds preceded the maturation of the conceptual ones, in the Soviet case theoretical activity preceded technological procurement and combat experience. The intellectual history of this innovation suggests that cultural, ideational, institutional and personal factors significantly conditioned the development of modern military theory.[1] This book aims to reflect these complexities and to explore contemporary military thought and innovation from various angles, focusing on the difficult balance between *anticipation* and *adaptation* in these matters.

In the opening chapter Azar Gat situates the IT-RMA discussion in the broader historical context of the technology-driven military innovations of the industrial age. The chapter discusses three revolutionary waves of the civil–military technological change which 'have been nothing less than the defining developments of modernity'. The First Industrial Revolution, which centered on the steam engine and on major advances in metallurgy, machine tools, and communications, profoundly increased strategic mobility and generated a revolution in firearms. The Second Industrial Revolution was dominated by chemicals, electric power and the internal combustion engine, which introduced tactical mobility and armored protection on to the battlefield. Finally, Gat's survey covers the latest revolutionary breakthroughs of the twentieth century – the Nuclear and the Information Technology Revolutions; he also explores the emerging revolution in WMD, specifically devoting attention to biotechnology. Gat identifies traits and fundamental problems that are common to all these revolutions. Transformations in military regimes are a dialectical phenomena: 'force multipliers' which produce one-sided battlefield results are usually matched by 'canceling out-effects' as rivals adopt countermeasures. The putative 'growing lethality' of military technology is balanced by 'exponentially increased protective power'. The growing weight of advanced hardware does not become prohibitively expensive. While one can only be more or less successful in predicting the contours of the future, there is a real-time need to work out the exact practical implications of the expected changes in war and to 'devise concrete programs of transformation in the organization and doctrine of the armed forces.' Gat's chapter highlights the complex and nuanced nature of revolutions in military affairs, and explains why armed forces are sometimes 'inclined to prepare for the last, rather than the next, war'.

What is the role of military doctrine in the era of wartime adaptation and bottom-up learning? What does defense transformation mean for the experts charged with producing military doctrines? How is military doctrine to be kept elastic enough to be able to address the different domains of warfare and endless scope of missions without deteriorating into parsimonious definitions that will prove banal, when theory meets practice? Is military doctrine, as a conceptual tool that organizes behavior, a relic of the past? In his chapter, Harald Høiback deals with these practical questions while situating them in the broader context of strategic theory.

If one accepts Plekhanov's claim about the role of the individual in history, the name most closely associated with the IT-RMA would probably be that of Andrew W. Marshall. Stephen Rosen discusses the impact of the Office on Net Assessment, headed by Mr. Marshall, on the ideas which dominated strategic studies and professional military discourse in the U.S. and worldwide in the last decades. Rosen demonstrates the role that Mr. Marshall played, along with Albert Wohlstetter, in the course of this military innovation, and explains why Mr. Marshall's thinking was more comprehensive than that of the Soviet theoreticians who inspired him, and broader than that of many in the American defense milieu. Based on generally available material and his own impressions, Rosen's chapter outlines the dominant American perspective on the technological changes, the particular input brought to this matter by Mr. Marshall, and describes 'what the Office of Net Assessment did to try to promote thinking by the

American military on this subject, and what the American military did in response'.

In Chapter 5, Antulio J. Echevarria outlines the most important developments in American military theory prior to the terrorist attacks of September 11, 2001. The American military is often perceived as being practically minded with only a limited interest in theoretical approaches. Echevarria, however, argues that there exists an American military tradition emphasizing the importance of operational theory and doctrine. He then discusses the complex interaction between operational experiences and the evolution of such theories and doctrines. With operation Desert Storm in the first Gulf War as a point of departure, he draws attention to the institutional aspects of this interaction in the 1990s. Inter-service rivalry proved to be a strong driving force behind the development in theory and doctrine during this decade, resulting in different schools of thought emerging within the American military. In particular the interpretation of operational experiences varied significantly between the Army and the Air Force.

The IT-RMA stands out in terms of its depth, pace and scope of diffusion. This is particularly impressive given the traditional conceptual conservatism of military organizations. The U.S. RMA terminology was emulated by military organizations worldwide and created a normative view of a modern conventional military. In Europe, Asia, the Middle East and Russia, military professionalism became associated with the adaptation of, or at least acquaintance with, the RMA school of thought. This book demonstrates this trend through two very contrasting case studies: diffusion of the military theory to the professional community of a superpower – the case of China; and the emulation of the RMA by the small state military – the case of Norway.

Jacqueline Newmyer Deal analyzes contemporary Chinese variations on the RMA theme. Her chapter seeks to capture the essence of modern Chinese military thought, identifies its main postulates and discusses the unique lexicon of the Chinese RMA. Newmyer Deal refers to sources of inspiration, intellectual foundations, strategic tradition and frames of reference informing Chinese military theoreticians. Relying on the rich collection of primary sources, she distils the unique characteristics of the Chinese RMA that distinguishes it from other innovations under a similar heading. Newmyer Deal puts forward a critical analysis of Chinese military thought and estimates the impact of the Chinese RMA on the correlation of forces between Bejing and Washington.

Military organizations often copy colleagues from abroad that they subjectively perceive as successful and victorious. These emulations are often conducted in an uncritical manner, so that the innovation does not necessarily fit the operational requirements and cultural environment of the receiving nation. Kjell Inge Bjerga and Torunn Laugen Haaland identify this pathology by taking a closer look at the Norwegian Armed Forces since the 1990s. They found that doctrinal evolution in Norway was detached from combat experience and operational necessities; it was mainly shaped by concepts inspired by the Great Powers, specifically by the U.S. RMA-based ideas. Bjerga's and Haaland's findings resonate with a history of uncritical emulation of U.S. practices by the Israeli Defense Forces, recently discussed by several scholars, and they suggest the existence of an intriguing pattern.

Contemporary military thought and innovation is a child of several epochs. Using Rosen's classic terminology, the IT-RMA concept was born as a 'wartime innovation' in the context of symmetrical–conventional Cold War contingency. It was boosted intellectually as a 'peace time innovation', facing a somewhat different set of operational challenges.[3] Eventually the RMA ideas were tested and modified as a 'transformation under fire', but against an enemy of a completely different nature. Focusing on the enemy's nature, the chapter by Itai Brun and Carmit Valensi brings to the research agenda an under-discussed phenomenon of the 'Other RMA'. Brun and Valensi argue that while military theoreticians on both sides of the Atlantic engaged in producing revolutionary concepts, a parallel school of military thought originated within several state and non-state strategic communities in the Middle East. What scholars tend to see today as the theory and practice of asymmetrical or hybrid warfare and as a countermeasure to the IT-RMA, Brun and Valensi dub the 'Other RMA'. Utilizing primary sources in Arabic and Farsi, they argue that this parallel innovation has its own anthology. They uncover its intellectual sources and demonstrate how these two streams of military thought, the IT-RMA and the O-RMA, have engaged each other operationally in actual combat since the late 1990s.

Indeed, most of the IT-RMA militaries found themselves on battlefields that did not fit the visions of the idealized RMA and were forced to significantly modify their concepts of operations and adjust them to current security environments. Historically, most of the defense transformation ideas were generated deductively and were disseminated in a top-down manner. However, when this innovation turned into 'transformation under fire', significant intellectual energy and conceptual insights originated from the operational echelons on the ground. Utilizing unique primary sources, Theo Farrell and James Russell look into the interplay of RMA theory with actual operational practice, and examine how the bottom-up (battlefield) wartime learning process significantly informs a professional approach to warfare.

Theo Farrell shows that military innovation is not always a necessary factor for an improvement of operational performance or for a major victory in war. Analyzing British combat experience in Afghanistan from 2006–2009, Farrell distils a more definite requirement for operational improvement – *military adaptation.* He separates this concept from the term *innovation* and suggests *military adaptation* as a separate theory for scholars of strategic studies. James Russell carries out a similar theoretical mission on a different empirical battlefield. He uncovers the sources of adaptation and innovation of the American forces by exploring their recent counterinsurgency experience in two Iraqi provinces. He found that the lion's share of the transformation of the tactical units was informed by organic, ground-up learning, which was not always in accord with the initial, top-down doctrinal expectations.

The critique of the IT-RMA became an essential driver for the development of the modern military theory. Current theoretical debates go beyond initial discussions of whether the recent transformation in the nature of warfare represents a revolutionary or an evolutionary discontinuity in military affairs, an issue debated vigorously following the publication of Stephen Biddle's renowned *Military Power*.[4] Today the critiques blame the proponents of the RMA for self-delusion, arguing that they outlined the preferred way of war and then assumed that the preference was relevant. Utilizing empirical evidence from the recent American campaigns in Iraq and Afghanistan, Biddle shows in his chapter how ideas of defense transformation face difficulties during the most frequent types of modern warfare (counterinsurgency, and stability and support operations), arguing that the RMA thesis might be invalid even for major conventional combat.

While Biddle bases his criticism on an analysis of recent operational evidence, Rolf Hobson proposes a critique from the perspective of a scholar of the RMA's intellectual history. In the early 1990s, argues Hobson, American theoreticians of the IT-RMA enquired into the historical examples of revolutionary innovations in military affairs, to inform their thinking about the emerging military regime. The examples from inter-war innovations in Europe, particularly the case of the German Blitzkrieg doctrine, fascinated them, and the lessons they learned improved their ability to conceptualize the RMA that was under way. Dealing with the case of Blitzkrieg as a frame of reference for the American theoreticians in the 1990s, Hobson argues that from the outset this example was based on outdated or irrelevant historical interpretations. Hobson refers the reader to the growing wave of revisionist literature on Blitzkrieg that devalues this supposed doctrine as a credible example of an historical RMA. Without necessarily disqualifying the IT-RMA thesis per se, Hobson argues that the most important historical precedent it cites is in fact an illustration created in its own image.

Andrew W. Marshall, the luminary of the American defense milieu, repeatedly emphasized the intellectual challenge associated with the development of military thought. Foreseeing an RMA is not a talisman for military victory, but the sooner defense experts recognize the discontinuity in military affairs, the better. The price of delay can vary from tactical–operational ineffectiveness to strategic catastrophe with devastating consequences for national security.[5] Andrei Kokoshin, one of the leading contemporary Russian strategic thinkers, echoes Marshall's view in his works on military innovations.[6] Extending this idea, two distinguished scholars of war, Azar Gat and Lawrence Freedman, have argued that as we are discussing the intellectual history of modern military thought, a new military regime might be emerging.[7] Inspired by the dicta of these renowned strategic scholars, the multifaceted inquiry presented in this book suggests a following insight for the students and practitioners of strategic affairs. Winning the next war equally requires *anticipation* of the emerging military regime, conceptual, technological, and organizational *innovation* associated with the new strategic environment, and an aptitude for *adaptation*, when operational reality on the ground challenges what Stephen Rosen defines as a new theory of victory.[8]

Notes

1 Dima Adamsky, *The Culture of Military Innovation: The Impact of Cultural Factors on the Revolution in Military Affairs in Russia, the U.S. and Israel*, Stanford CA.: Stanford UP, 2010.
2 Avi Kober, 'The Israeli Defense Forces in the Second Lebanon War: Why the Poor Performance?', *Journal of Strategic Studies*, 31: 1, 2008, pp. 3–40.
3 Distinction between *war time* and *peace time* innovation is taken from Stephen Peter Rosen, *Winning the Next War: Innovation and the Modern Military*, New York: Cornell University Press, 1991.
4 See: 'Military Power: A Roundtable Review', *Journal of Strategic Studies*, 28: 3, 2005, pp. 413–69.
5 Adamsky, *The Culture of Military Innovation*, pp. 1–2.
6 Andrei A. Kokoshin, *Innovatcionnye vooruzhennye sily i revoliutciia v voennom dele*, Moscow: URSS, 2009; Andrei A. Kokoshin, *O revoliutcii v voennom dele v proshlom i nastoiaschem*; Moscow: URSS, 2006, p. 36.
7 Remarks by Azar Gat and Lawrence Freedman, concluding roundtable of the international conference 'Modern Military Theory – a Critical Examination' (hosted by the Norwegian Institute for Defence Studies, Oslo, 24–25 June, 2009).
8 The term is taken from Rosen, *Winning the Next War.*

2 The 'revolution in military affairs' (RMA) as an analytical tool for the interpretation of military history

Azar Gat

The conspicuous changes that have taken place in the face of warfare over the past decades have been titled: the 'Revolution in Military Affairs' (RMA). The problem with this label, however, is that it tells us nothing about the nature of the revolution and its place in the broader context of technology-driven revolutions of the industrial–technological age. These have been nothing less than the defining developments of modernity. Over the past two centuries innovation in technology accelerated dramatically in comparison to pre-industrial times, with military technology constituting merely one aspect of a general trend.

In pre-modern times, too, technology mattered, and some innovations in military technology profoundly affected warfare and history in general. Metal weapons, equestrian technology, the longbow, the rowing and sailing ships, and firearms are oft-cited examples, and there are many more. And yet, technology improved slowly in the pre-modern era, so that something close to equilibrium often prevailed for millennia between each significant 'punctuation' in the evolution of military technology. The main infantry weapon, the musket, changed little between 1690 and 1820. However, from the beginning of the industrial–technological era, as military theorist J. F. C. Fuller saw, the pace of technological innovation became such that the best armed force of one generation would have been totally unable to confront in the open a well-equipped opponent of the following generation.

As Fuller equally saw, the advances in military technology were closely related to civilian developments; neither took place evenly over time nor across the technological front, but were mainly clustered around consecutive breakthroughs in a number of sectors each time.[1] Taking decades to run their course, these technological breakthroughs then gave way to other breakthroughs in different sectors. Although some oversimplification is necessarily involved, Fuller rightly identified three such major revolutionary waves of civil–military technological change during the nineteenth and twentieth centuries.

Three revolutionary technological waves

The so-called First Industrial Revolution, taking about a century to unravel, centered on the steam engine and on major advances in metallurgy and machine tools. The steam engine, practically the only engine in existence until the late nineteenth century, was applied to propel all sorts of different machines, revolutionizing one field of human activity after another. Originally developed to pump out water from mines, it was then harnessed to the newly developed spinning and weaving machines of the cotton industry, revolutionizing textile production. Applied to pull trains of wheeled carriages that ran on railroads, it revolutionized land transportation from the 1820s on, placing it for the first time on equal footing with water transportation and opening up the interior of the world's great continental landmasses. And harnessed to a paddlewheel and then a propeller, it finally displaced the great sailing ships, one of the pinnacles of pre-mechanized technology. All these were various applications of the same basic technology.

The above changes affected the military field as deeply as they did civilian life. The railway increased armies' strategic mobility and logistical capability by a factor of hundreds. While naval mobility only doubled or tripled as steam replaced sail, naval tonnage grew four or fivefold and (iron and steel) battleships' size – and might – tenfold and more.[2] To these was added the revolution in information communications, as electric telegraph lines connected not only armies across countries but also allowed naval bases spanning oceans and continents to communicate in real time, where once weeks, months, and years had been necessary.

Simultaneously, during the nineteenth century, the revolution in metallurgy (iron followed by steel) and machine tools generated a revolution in firearms and tactics. Rifling and breech-loading were pioneered in infantry firearms during the 1840s, and in artillery during the 1850s and 1860s. Magazine-fed rifles, 'repeaters', were developed in the 1860s and 1870s, and quick-firing artillery, using a hydraulic mechanism to absorb the gun's recoil, in the 1880s and 1890s. In consequence, range, accuracy, and rapidity of fire each increased some tenfold within 60 years, not counting the development of the automatic machinegun from the 1880s, which multiplied firepower yet more.[3] Naval gunnery underwent similar developments, to which the torpedo was added from the 1870s.

All these, however, were lopsided revolutions, especially on land. As in the economy so in the military, spheres of activity to which the steam engine could not be applied remained manual and unaffected by the Revolution. Thus, while armies rode trains on their way to the battlefield and were easily controlled by telegraph, they fell from the pinnacle of high-tech communications back to Napoleonic if not Alexandrian times once *on* the battlefield. Their campaign and tactical mobility remained confined to human muscles, with their artillery and supplies drawn by horses. Millions of horses remained in each of the Great Powers' armies during World War I, and in some, including the mythically mechanized German army, also throughout World War II. Field command and control, where telegraph lines could not be laid in advance, was similarly downgraded to messengers on foot or horseback. Furthermore, whereas firepower increased tenfold and more, troops, while dispersing and taking cover, still had nothing better than their skin to protect them from the storm of steel on the open field. Hence the murderous stalemate on the Western Front during World War I, both tactical and operational. Even those puny gains made by attacking infantry at terrific cost were reversed as decimated foot soldiers, struggling to extend their tactical gains deeper, were pushed back by enemy reinforcements rushed up by rail.

However, from the 1880s a new revolutionary wave of industrial technology, the so-called Second Industrial Revolution, was beginning to unravel in civilian life, affecting the military field as profoundly as the First Industrial Revolution had. Chemicals, electric power, and the internal combustion engine dominated that second revolutionary wave. While the chemical industry contributed high explosives – remember Alfred Nobel – and was soon to produce chemical warfare, and while developments in electricity also had various military applications, including radio communication, it was the internal combustion engine that affected war the most decisively. Lighter and more flexible than the steam engine, it made possible mobility in the open country, away from railways. Passenger and transport automobiles (as well as the tractor) rapidly evolved between 1895 and 1905, increasing cross-country mobility by a factor of tens. World War I inaugurated the tank – an armored and armed tractor – which introduced mechanized mobility and mechanized armored protection into the battlefield, thereby redressing the huge imbalance created by steam. Controlled by radio, which similarly extended real time information communication on to the open field and away from stationary telegraph lines, mechanized armies on tracks and wheels had matured by World War II, some half a century after the pioneering of the technologies that had made them possible.

Simultaneously the internal combustion engine also made possible mechanized air flight. A remarkably similar trajectory followed, with the first such flight taking place in 1903, and massive air forces quickly coming into being during World War I and further developing by World War II. Ships, already steam powered and armored, were less dramatically affected by the internal combustion engine. Nonetheless, naval warfare in general was revolutionized. Dual propulsion by the internal combustion and electric engines made possible the first workable submarine in 1900, while the aircraft was to bring about the demise of the gunned battleship. Thus, the automobile, the submarine, and the aircraft made their appearance in close proximity roughly between 1895 and 1905, all made possible by the same new technology. They all developed in leaps and bounds during

World War I, and together they completely dominated both land and naval warfare in World War II.

By then new technological breakthroughs were beginning to make their mark in other sectors, most notably electronics, which revolutionized both civilian life and war in the so-called Third Industrial or Information Revolution. Radar, developed in the late 1930s, deeply affected air, air–land, and sea warfare in the following decades. From around 1970, electro-optic, television, and laser guidance for missile weapon systems began to revolutionize air–land and land battle. Since then, sensors of all sorts have been rapidly improving, in connection with the fast miniaturizing micro-chip that has doubled electronic computation capacity every 18 months for nearly half a century. The microchip – the steam or internal combustion engine of our times – has been applied to a dazzling array of new technologies, revolutionizing each. As a result of all this the identification, acquisition, and destruction of most hardware targets has become almost a foregone conclusion, nearly irrespective of range. Showing little sign of leveling off, the electronic revolution is bringing about increasing automation. This is the electric–robotic warfare that the pioneering Fuller predicted as early as 1928 as the third great wave after mechanization.[4]

The far-reaching effects of the ongoing electronic–information revolution on warfare have been endlessly discussed. The old mechanized–armoured armies of the previous era may not be disappearing, but they have been shrinking in size and transformed to embrace electronic warfare themselves, defensively as well as offensively. The two Gulf wars demonstrated this most strikingly, for the Iraqi side that lacked the new technologies found out to its cost that its numerous old-style formations were as vulnerable as herds of prehistoric mammoths. The gap between developed and less developed protagonists seems to have widened considerably. And yet the latter have been adjusting more quickly, and in ways different than expected.

In the first place, less developed players have been moving to get rid of their heavy formations, adopting instead low-signature troops, weapons, and tactics. They aim to slip under the radar of the electronic weapon systems, which are much better at identifying hardware than people. (Notably, though, the weaker side cannot dispense with its heavy dual-purpose civilian–military infrastructure that remains highly vulnerable, as demonstrated by Serbia's experience in the Kosovo War.) Second, the massive market penetration of new technologies into every aspect of daily life makes them available to less developed players as well, perhaps not in the form of the most expensive cutting-edge military systems but more as widely-available and cheap gadgets. Satellite navigation systems (GPS) that offer precision guidance, computer networks that can be exploited and disrupted, and cellular phones that can be activated from afar, are some examples. Indeed, high-tech technologies have both polarized *and* democratized the balance between the more and less advanced sides in war, for the means to generate massive damage with pinpoint accuracy has been trickling down below the state level, becoming available to non-state actors as well.

Common traits and fundamental problems

This is taking us from the survey of history to discuss some general, recurring traits of military–technological revolutions:

Force multipliers, one-sided battlefield outcomes, and canceling-out effects

With arms races gaining a wholly new significance in the technological age, swift changes in and redresses of the balance of power due to innovation have become very much the rule. Particularly when one side succeeds in securing a decisive lead in the acquisition and assimilation of breakthrough weapon systems, it might gain a 'force multiplier' that can produce one-sided battlefield results. Examples include the Prussian breech-loading rifle in the 1866 war against Austria, the German mechanized forces at the beginning of World War II, and the American electronic weapon systems in the campaigns of the past two decades. Over time, however, there would be a canceling-out effect, as rivals caught up in the development and assimilation of new weapon systems and operational doctrines, and adopted countermeasures. This, too, was pointed out early by Fuller. Breech-loaders responded to breech-loaders; mechanized forces and anti-tank weapons countered mechanized forces; electronic disruption and suppression devices targeted electronic weapon systems.

Technology could also be transferred – sold or given – which occasionally narrowed the gap in power between more and less advanced rivals. Finally, extreme technological asymmetry has been countered by asymmetrical strategy. The Battle of Omdurman (1898) in the Sudan, where Kitchener's British army, employing machineguns and magazine-fed rifles, annihilated the Dervish army, killing 11,000 at the price of 140 British casualties, demonstrated the huge gap that had been opening between the armies of technologically advanced powers and their rivals from less developed societies. Thereafter, the latter tended to avoid direct fighting, opting for guerrilla and other methods of irregular warfare. At sea and in the air, where fighting has been the most hardware-intensive and has been carried out entirely by weapon systems (rather than by armed men), the gap between technologically more and less advanced rivals has been the widest.

Growing lethality?

Intertwined as it has been with the general development of technology, weapon systems' effectiveness, while difficult to measure exactly, probably increased at the same rate as productivity in general during the industrial–technological age: that is, by a factor of hundreds. But what exactly does this increase signify? Destructiveness or lethality may appear to be the issue, for that is what war is all about. On the other hand, as the above survey has already demonstrated, developments in military technology also exponentially increased protective power – for example, through mechanized defensive armor at sea and on land, through growing (indeed sometimes literally rocketing) swiftness and agility, and through electronic counter-warfare. Contrary to widespread assumptions, studies of war lethality, measured by military and civilian casualties, show no significant increase during the nineteenth and twentieth centuries in comparison to earlier historical periods, relative to population.[5] Mortality as a percentage of the armed forces did not grow.

Furthermore, the vast majority of the non-combatants killed by Germany during World War II, the most severe modern war – Jews, Soviet prisoners of war, Polish and Soviet civilians (millions in each category) – fell victim to intentional starvation, disease, and mass executions, rather than to any sophisticated military technology. Indeed, similar and greater rates of combat mortality than that experienced in World War II, had been commonplace among pre-state, small-scale societies that used the most primitive military technology. Instances of genocide in general during the twentieth century, as earlier in history, were carried out with the simplest of technologies, with Rwanda being a chilling recent example.

Spiraling costs?

Another general characteristic of the industrial–technological age, with its consecutive military revolutions, has been the steadily growing weight of spending on military hardware. However, again contrary to a widely held view, weapon systems' cost, like military spending in general, did not spiral out of control with every new generation, nor did they become prohibitively expensive. Military spending does not simply rise with a 'free-floating' cost of weapons, but is always bounded by economic capabilities, by priorities, and by an enemy's spending bounded by similar constraints. In fact, the peacetime military spending of the powers during the industrial age has remained remarkably stable, at less than 5 percent of GDP. That was the level of military spending in Britain, the leading industrial power of the nineteenth century, as well as in the other European Great Powers. Although registering some increase, military spending retained these modest levels even during the arms race between the European powers in the decade before World War I.[6] During the twentieth century peacetime military spending generally retained the same, surprisingly resilient, levels as in earlier centuries: over 5 percent of GDP for the United States during the Cold War; around 3 percent for the European countries. Thus, military spending did not 'spiral', but generally kept in line with the overall growth in GDP. Only in wartime, most notably during the two world wars, did the military spending of the warring Great Powers leap to around half of total GDP.

What is true, however, is that hardware's *share* in military spending has grown during the industrial–technological era, with the rising cost of some weapon systems compensated for by smaller numbers or by a changing balance within military procurement, both matters of cost-effectiveness.[7] The problems of priorities that the American defense authorities are facing today with such expensive weapon systems as the air superiority aircraft F-22 Raptor are not new.

During the early modern period, pay and provisions for the men far outweighed the cost of military hardware, not only in the armies but also in the much more capital-intensive navies. However, as the armed forces became increasingly mechanized and hardware-dependent, this ratio tilted in the other direction. In the navies, the first intensively mechanized service, manpower costs (including pay, provisions, and clothing) declined to about 40 percent of the naval budgets by the last decade of the nineteenth century, and to around 30 percent by the first decade of the twentieth century. At the same time warship construction costs grew to over one third of naval budgets, with nearly as much additional spending dedicated to repairs, maintenance, and ammunition. Thus materiel grew to outweigh manpower costs by two to one.[8] The U.S. Navy's budgetary figures a century later, in the 2000s, remain in the same range for manpower, procurement plus R&D, and operations and maintenance, respectively.[9]

Armies were far less mechanized than navies around 1900, so manpower costs (direct and indirect, where conscription was involved) undoubtedly still constituted the majority of army expenditure in all the Great Powers. Yet this picture is misleading, because one of the major elements of military power was the railroad system – for mobilization, strategic mobility, and logistics. And the overwhelmingly civilian infrastructural costs of that system (and others) did not register in military budgets. From World War One onwards, however, armies increasingly mechanized. Consequently, manpower costs in the U.S. Army, for example, fell to around 40 percent of the Army's budget in the 2000s, even though this is an all professional rather than a conscript force. Materiel costs (excluding construction) amount to about half of the budget, with operations and maintenance costing double, as much as new procurement and R&D.[10]

Finally, while armies remain the most labor-intensive armed service, air forces are the most capital-intensive of all armed services. Manpower costs amount to only 20 percent of the U.S. Air Force's budget in the 2000s, with operations and maintenance, procurement, and R&D together taking up about three-quarters of the budget.[11]

Prediction under conditions of discontinuous experience

The fundamental challenge posed by military–technological revolutions is that it involves a sweeping change that goes far beyond existing experience. The contours of the change are shrouded in the mists of the future, and can only be more or less successfully predicted. Nonetheless, huge investments in hardware and far-reaching adjustments in military structure and operational doctrine, which can only be tested in battle, are called for in the present. To be sure, the uncertainty of the future is not confined to the military sphere. Firms, states, and individuals are required to make decisions on future actions and investments based on predictions.

Still, the military sphere is special in that the practical experience of war is not continuous but is usually limited to relatively short spates of active belligerency, separated by long periods of peace. Although armed forces attempt to overcome these gaps by means of operational analyses, simulations, and exercises of all sorts, a continuous reality check for the military does not exist in the same way that it does, say, in economic life. Granted, the limitations of reality check in the economy, too, have recently been most poignantly demonstrated by the buildup, and final bursting in 2007–2008, of the credit bubble economy, with the mountain of economic assumptions and expectations that went with it. But the limitations are even more severe where experience is not continuous.

In Europe, for example, peace prevailed among the Great Powers for 43 years, from 1871 to 1914. Wars took place only on the peripheries of the system. As a result, even though the exponential rise in firepower during that period was well recognized by all armies, its significance was not fully assimilated until the actual test of fire in World War I. The long interruptions in the experience of war is the reason why armed forces are proverbially inclined to prepare for the last, rather than the next, war, a problem that, indeed, became most acute only with the revolutionary technological changes of the industrial age.

I assume most of us would have gladly traded combat experience for peace. Unfortunately, however, in recent years the American and Israeli armed forces have gained continuous combat experience, with the result that gaps in the assessment of the use of modern weapon systems on the electronic battlefield narrow, and continuous adjustments are carried out. Much of this experience, however, is limited to counterinsurgency warfare and to regular combat against far weaker rivals. And limited experience in specialized circumstances can be more misleading than no experience at all. The lessons of the Spanish Civil War, for example, retracted the development of independent mechanized forces, most notably in the Soviet Union.

Grand visions and practical designs

It may not be redundant to point out that the problem in dealing with technological revolutions is not so much to realize generally that such a revolution is in fact occurring, but to work out its exact form and practical implications. While inspiration is crucial, god has long been proclaimed to be in the details. Insufficient awareness of a brewing revolution may exist in its formative stage; but awareness grows quickly as the revolution gathers momentum, and all the more so in our technological society, which has grown accustomed to such changes and experiences them in all spheres of life. The real problem is to identify the precise nature of the expected changes in war, devise concrete programs of transformation in the organization and doctrine of the armed forces, and determine the optimal mix of hardware to be developed and procured. All these complex and critical questions are interconnected, and they are saturated with technological and other unknowns, not to mention budgetary and operational constraints. There were many early car manufacturers, but only Henry Ford got it exactly right, at least for a while. There have been thousands upon thousands of high-tech companies, but only Microsoft, Apple and Google developed winning combinations.

During the 1930s, for example, contrary to mythology, the debate within the armed forces did not take place between far-sighted exponents of mechanization and their reactionary rivals, but revolved around far more practical questions: should all tanks be concentrated in armored divisions, or should some be allocated to infantry support? And, following on from this, what types of tanks should be manufactured, and what should be the role of infantry and other traditional arms on the modern battlefield?[12] Conclusive answers to these questions were only forthcoming following the practical test offered by World War Two, and not always along the lines suggested by the radicals.

This applied even more to air forces. Their major role in any future war was recognized by all, and it was in fact the radicals who were proved mistaken in their most central assumptions. They believed that air forces were capable of easily and swiftly destroying the industrial infrastructure of Great Powers and force them to capitulate. They thus contended that armies and navies had only a secondary role to play, if they had not become wholly obsolete already. At sea, it was not clear, for example, if the invention of ASDIC/sonar spelled the end of the submarine, as Winston Churchill, for one, had hoped. Nor was it known if the aircraft had rendered the battleship obsolete or would co-exist with it. These critical questions were tested, and settled, only in World War II.

The nuclear and biotechnological revolution

Our discussion so far has been confined to conventional warfare. Obviously, though, by far the most revolutionary technological breakthrough of the modern era has been the advent of nuclear power and nuclear weapons. The overwhelming destructive power of a nuclear explosion is offset by no parallel rise in defensive power, as has been the case with other military technologies. This is finally the ultimate weapon, a doomsday machine that gives anybody who possesses the necessary stockpile an assured ability to completely destroy his enemy – indeed, the world. At the same time the fact that no effective defence against nuclear weapons exists means that when both sides possess them their actual use in war has been all but prevented by deterrence of supreme potency: the promise of mutually assured destruction (MAD). This was the logic that dominated the Cold War and served as the ultimate break against it becoming hot. And yet this logic may no longer apply as comprehensively as it did.

The change is driven in part by technological innovation and in part by technological proliferation. It concerns not only nuclear but also biological weapons. Nuclear technology has been around since 1945. But as both its civilian and military uses are becoming more widely available, the Non Proliferation Treaty regime built to contain it is in real danger of collapse. Biological weapons have also been around since before World War II. Yet with the breakthroughs achieved in decoding the genome and genetic engineering, the biotechnological sector has become one of the spearheads of today's scientific–commercial revolutionary wave, together with electronics. As a result biological weapons have become both much more lethal and accessible. A virulent laboratory-cultivated strain of bacteria or virus, let alone a specially engineered 'superbug' with no immunological or medical antidote,[13] could make biological weapons as lethal as nuclear attacks. At the same time they are far more easily available than nuclear weapons.

By the late 1990s there were already an estimated 1,300 biotechnology companies in the United States and 580 in Europe.[14] Since then the number of scientists authorized to work with biological agents has risen to 15,000 in the U.S. alone, rendering monitoring by government agencies all the more difficult.[15] One estimate suggests there are some 20,000 labs in the world, where, within the next decade, a single person will be able to synthesize any existing virus. In the same labs, five people with $2 million will be able to create an enhanced pathogen – a virus that could infect and kill people who have been immunized with conventional vaccines. With $5 million the same five people could build a lab from scratch, using equipment purchased online.[16]

The root problem in this process is the trickling down to below the state level of the technologies and materials of WMD. The ability of non-state actors to buy, steal, rob, and/or manufacture WMD has increased dramatically. Here lies the bewildering nature of the new mix that is unconventional terror. Not only are terror groups more likely to consist of zealots who are willing to sacrifice their lives and may positively desire a general apocalypse, they are also too elusive to offer a clear enough target for retaliation, on which the whole concept of deterrence is based. For this reason the use of ultimate weapons is *more* likely to come from them than it is from states, even though the latter possess far greater capability. In contrast to the mindset that dominated strategic thinking since the onset of the nuclear age, unconventional capability acquired by terrorists is *useable.*

The 'encapsulation' of destructive power in WMD, particularly the nuclear and biological, creates a situation whereby one no longer has to be big in order to deliver a devastating punch. Scenarios of world-threatening individuals and organizations, previously reserved to fiction of the James Bond genre, suddenly become real. The greatest threat of nuclear and biological proliferation into countries with low security standards and high levels of corruption is not so much that of actual military use by the states involved but the far-increased danger of leakage to non-states. Not only might people and organizations with access to nuclear facilities sell or otherwise transfer nuclear materials, expertise, and even weapons to others, states in the less developed and unstable parts of the world are also ever in danger of disintegration and anarchy. Abdul Qadeer Khan, the Pakistani nuclear scientist who headed his country's program to manufacture an atomic bomb, sold the nuclear secrets to perhaps a dozen countries, reportedly for as little as millions or tens of millions of dollars. And the prospect of a Taliban takeover, or general anarchy, in Pakistan, is looming large even as these lines are written, to the concern of the world community. The disintegration of the Soviet Union left as part of the debris of its advanced military infrastructure unemployed scientists, poorly guarded production facilities, unaccounted for materials, and, most troubling of all, the weapons themselves. Indeed, the collapsed Soviet Union rather than the former nuclear superpower may be the model for future threats. In the vast and barely reachable parts of the globe that are inhabited by fragmented and unruly societies, the ability to monitor unconventional weapon production or acquisition is inherently limited. The difficulty of finding a needle in a haystack is pale by comparison.

According to Rolf Mowatt-Larssen, the U.S. Energy Department's director of intelligence, al-Qaeda was seeking to acquire weapons of mass destruction for nearly a decade before September 11. Osama bin Laden offered $1.5 million to buy uranium for a nuclear device. In August 2001, just before the attack on the U.S., bin Laden and his deputy, Ayman al-Zawahiri, met with Pakistani scientists to discuss how al-Qaeda could build such a device. al-Qaeda also had an aggressive anthrax program that was discovered in December 2001, after the organization was driven from Afghanistan. In 2003, Saudi al-Qaeda operatives tried to buy three Russian nuclear bombs. Around the same time, Zawahiri decided to cancel a cyanide attack on the New York subway system, telling the plotters to stand down because 'we have something better in mind'. After 2004, al-Qaeda's WMD trail went cold.[17] The price of al-Qaeda's successful terror attack on

9/11 may have been to nip its far more dangerous plans in the bud. It seems clear that the loss of the Afghani safe haven was a serious blow to al-Qaeda's WMD program, though, of course, the price of success is that we shall probably never know for certain. The failure of al-Qaeda to initiate further terrorist acts on U.S. soil after 9/11 (pointed out by those who believe that the danger has been exaggerated) is largely attributable to the loss of that safe haven. As with other effective countermeasures, the paradoxical conspicuousness of the 'dog that didn't bark', in Sherlock Holmes's words, should not become a cause for complacency.

Notes

1 These ideas repeatedly occur in J. F. C. Fuller's voluminous writings; but see especially his *On Future Warfare*, London: Praed, 1928; and, *Armament and History*, London: Eyre, 1946.

2 Data for Britain. See Quincy Wright, *A Study of War*, Chicago: University of Chicago, 1965, pp. 670–1 (military); and B. R. Mitchell, *International Historical Statistics, Europe 1750–1988*, New York: Stockton, 1992, tables F4 (merchant).

3 Of the many references to these developments, the most expert are: Dennis Showalter, *Rifles and Railroads: Soldiers, Technology and the Unification of Germany*, Hamden, Conn.: Archon, 1975; and Daniel Headrick, *The Tools of Empire: Technology and European Imperialism in the Nineteenth Century*, New York: Oxford UP, 1981.

4 J. F. C. Fuller, *Towards Armageddon*, London: Dickson, 1937, pp. 92, 132.

5 The general trend is more or less agreed upon by statistical studies such as: Melvin Small and David Singer, *Resort to Arms: International and Civil Wars, 1816–1980*, Beverly Hills: SAGE, 1982; Jack Levy, *War in the Modern Great Power System, 1495–1975*, Lexington: UP of Kentucky, 1983; Evan Luard, *War in International Society*, London: Tauris, 1986. Niall Ferguson, *The Cash Nexus: Money and Power in the Modern World, 1700–2000*, New York: Basic Books, 2001, pp. 33–6, makes the point for spiraling lethality, but he scarcely controls for population (also as a function of the geographical scope of wars) and mobilization levels, and appears to be unaware of the heightened protective aspect of improved military technology.

6 David Stevenson, *Armament and the Coming of War: Europe, 1904–1914*, Oxford: Oxford UP, 1996.

7 Ferguson, *The Cash Nexus*, pp. 30–3; although I differ with some of his views here, our general conclusion is similar.

8 Jon Sumida, *In Defence of Naval Supremacy: Finance, Technology and British Naval Policy*, Boston: Unwin, 1989, Tables 3–14; Stevenson, *Armament and the Coming of War*, pp. 7–8.

9 U.S. Department of Defense, *National Defense Budget Estimates for FY 2004* (March 2003), pp. 154–9, 172–7, 190–5 (available online).

10 Ibid., pp. 148–53, 166–71, 184–9.

11 Ibid., pp. 160–5, 178–83, 196–201.

12 Azar Gat, 'Ideology, National Policy, Technology and Strategic Doctrine between the World Wars', *Journal of Strategic Studies*, 24: 3, 2001, pp. 1–18.

13 Philip Cohen, 'A Terrifying Power', *New Scientist*, 10, Jan. 30, 1999; Rachel Nowak, 'Disaster in the Making', ibid., Jan. 13, 2001, pp. 4–5; Carina Dennis, 'The Bugs of War', *Nature*, May 17, 2001, pp. 232–5.

14 Nadine Gurr and Benjamin Cole, *The New Face of Terrorism: Threats from Weapons of Mass Destruction*, London: Tauris, 2000, p. 43.

15 Spencer Hsu, 'Modest Gains against Ever-Present Bioterrorism Threat', *Washington Post,* Aug. 3, 2008.
16 Anonymous scientist cited by Anne Applebaum, 'The Next Plague', *Washington Post,* Feb. 18, 2004.
17 David Ignatius, 'Portents of a Nuclear Al-Qaeda', *Washington Post,* Oct. 18, 2007.

3 What is doctrine?

Harald Høiback

When it comes to innovation, military forces seem to be of a dual nature. On the one hand, the armed forces are not the place to look if you want genuine novel thinking. Moreover, very little of what you find in the military inventory has its origin there:

> During the twentieth century, too, none of the most important devices that have transformed war – from the airplane through the tank, the jet engine, radar, the helicopter, the atom bomb, and so on all the way down to the electronic computer—owed its origins to a doctrinal requirement laid down by people in uniform, though it is fair to add that some inventors did have military experience.[1]

In other words, military doctrine is not a device to spur creativity and inventiveness. The other side of the dual nature mentioned above, is that *when* a clever idea occurs, no one is better at utilising it than military organisations:

> If the military thus represent an exceptionally unfavourable environment for the mysterious creative act popularly known as invention, exactly the opposite is true concerning the no less important process by which an idea is taken from the drawing board and transformed into operational hardware.... Once the potential usefulness of a new concept is recognized, no organization is better placed than the armed forces to guide its development and bring it to fruition.[2]

It could thus be argued that military doctrines turn out to be rather inept at encouraging inventiveness. However, once the inventions are made, military doctrines are apparently second to none in propelling these inventions into workable hardware and better ways of warfare.

In this chapter our aim is to analyse doctrine a bit more closely. What is a military doctrine, actually? I will argue that doctrine can be a lot of things, and that its ability to transform ideas into operational hardware is a rather unsung element of it. Moreover, as it turns out, there are many voices around urging us to place doctrine in the dustbin of history.

NATO defines doctrine as 'fundamental principles by which the military forces guide their actions in support of objectives. It is authoritative but requires judgement in application.'[3] To modern ears this may sound a bit archaic. What on earth can some solemnly endorsed principles do for military forces that have a virtually endless scope of missions? What can be said, apart from generalities and banalities that address all soldiers in all possible situations? The task of the doctrinaires seems both insurmountable and inadvisable: 'there is no such thing as a doctrine of war and [if] there were such a thing it would be bad'.[4]

There are ample warnings of a similar sort around: 'Knowing the enemy is the bedrock of the business of strategy: strategic theories, in comparison, are second order problems. To concentrate on doctrine before enemies is to put the theoretical cart before the actual horse – a double error.'[5] In other words; to write a generic doctrine, for all kinds of operations, in all kinds of coalitions, and against all kind of enemies, is a waste of time, in the best of circumstances.

What's more; doctrine is particularly bad in the type of operations that have concerned us most recently, i.e. counterinsurgency (COIN). To succeed in COIN you have to delegate and decentralise, and leave the most important decision-making to mid-ranking officers in the field. And according to Markus Mäder, the 'general perception of traditional COIN advocates within the [British] Army, [is] that doctrine would jeopardize these very qualities'.[6] The political and cultural complexities of COIN operations require officers who are able to play it by ear, in the sense that they pay greater attention to the situation at hand than to some generic prose written by a typing pool far from the hot spot.

Consequently, doctrine seems to be a relic of the past. It once made perfect sense, perhaps, and was an apt solution to recurring questions, as for instance how to fight the same kind of war over and over again, but not any more, not when we have to fight new enemies in exceedingly changing circumstances. What can doctrine offer armed forces that have their units 'scattered from Malta to Peshawar, from the Curragh to the Cape'?[7]

Before we sign and issue doctrine's death certificate, however, we ought to know the patient a bit better. It would be a pity to bury a creature still alive. Despite NATO's definition, 'doctrine' is still a rather fluffy concept with a number of definitions and connotations.[8] So, in order to write doctrine off as an irrelevant military tool, we have to agree on what doctrine actually is.

La doctrine

Doctrine's origin can be traced in different centuries and in different cultures, depending on the preferred definition. If you are not too demanding on behalf of the actual *codification* of doctrine, you will find doctrines at the root of both the Macedonian phalanx and the Roman legions. However, if you insist on *written* doctrines, you have to wait until the paper revolution that followed Gutenberg's invention. Moreover, if you are searching for the fountain of *modern* doctrines, you have to wait until after the French Revolution.

Quite commonly, Helmuth von Moltke is heralded as the first real doctrinarian. According to for instance General Sir Rupert Smith, Moltke was the true inventor of 'a coherent military doctrine' and his 1869 *Instructions* have, in spirit at least, 'inspired most operation manuals of Western militaries to this day'.[9] Albert Palazzo also gives Moltke the mixed blessing of being the inventor of military doctrine.[10] However, a case can be made that it was not the Prussians that invented modern military doctrine.

Presumably, during Moltke's era the word *doktrin* did not exist in the German military vernacular.[11] Despite the fact that you find the word used by Carl von Clausewitz in for instance English and Swedish translations of *Vom Kriege*, you would presumably search in vain for the word in his original German manuscripts.[12] Even the idea of having a template to fight by would be antithetical to the German way of war:

> Nothing could be further from the German military mentality than making a fetish out of a word or phrase. No rule of war could be universally valid for all times and places. *Kein Schema!* Was one of the most common admonitions of the German staff, along with contempt for *Patentlösungen*.[13]

It was those on the receiving end of Prussian strategic culture and military professionalism that had to do something radically to catch up. Hence, in order to find the roots of modern military doctrine, we have to pay nineteenth century France a visit.

Military history has seen scores of shattering defeats, of which France has had a substantial share. Here however, the defeat against Prussia in 1870 is the crucial one. If you just look at the numbers and available technology involved, you find little that can explain the French disaster. In fact, that should not come as a surprise because the outcome of modern wars can rarely be explained by numbers and materiel, despite the long traditions of doing exactly that.[14] The disaster of 1870 could not even be explained by differences in bravery and individual skills.

Nonetheless, the French understood that the catastrophe of 1870 had revealed something seriously wrong in the French Army. To do better next time, France had thus to do something to its army's collective ability to fight modern battles. Combat's mental domain, as well as its physical, had to be investigated. The invincible spirit of war associated with the great Napoleon had to be rekindled in the French forces. They needed something to infuse 'with life the whole body of an army, without which the most admirable texts and precepts enshrined in regulations are but dead bones and dry dust'.[15] Hence they not only needed a new formula for victory, but also something to energize and inflame willpower and to foster what later became known as *élan vital.*

To stand firm in the storm of steel that evolved on the battlefield in the second half of the nineteenth century would take the conviction of a crusader, spurred by the holy gospel, by *la doctrine.* A doctrine could presumably evoke conviction within the military sphere just as it did within the religious:

> It is necessary that they [our officers] should understand not only the letter but also the spirit of our regulations, so that they may be certain that these regulations are their most sure guide and one to be followed with sincere conviction and entire faith.[16]

In 1903, after more than 20 years maturation, 'doctrine' was definitely established by Ferdinand Foch as a distinct military concept, apart from its originally religious connotations.[17] After another decade, during the Great War, doctrine finally became a household concept in all the great armies.

Worthy of note, however; after the Russian Revolution, there erupted a heated doctrinal debate between Trotsky and some of his military subordinates. To what extent that discussion rested on a Russian or a Marxist tradition, independent of the French discourse, is unfortunately hard to answer without being in command of the Russian language and well versed in the labyrinths of Marxism. Hence, that is a question I have to leave to more gifted investigators.

So far we have had a quick glance at the origins of modern military doctrine. It was a French invention with a double purpose, both explaining how to win the next war and infuse the will to do so. However, we have not said much about what doctrine *is.*

The anatomy of military doctrine

In order to abstract the essentials of military doctrine we will take a short detour via the world of sports. The point is not to trivialise war, but that sport is, luckily, a much more familiar arena for most of us.

In team sports there are in principle two ways to make the most of the team. One way is to gather the best players available and position them according to their skills and competence. During the match the players use their judgement and abilities as far as they can in order to take advantage of the opportunities on the pitch, supported by erratic orders from the bench.

The second approach is to base the game on a permanent system and a particular way to play. Players are picked, trained and play each game in accordance with a scheme. This is the doctrinal approach. The aim of this approach is to make the total greater than the sum of the parts, to have the best team without having the best players, so to speak. By providing a conceptual compass and a mental counterweight the team can hopefully also handle more chaos and confusion than they can do without.

A recipe, such as a doctrine, that tells us how to play in order to win has to contain three basic elements. At the outset it needs some assumptions about what leads to victory. It needs an element of explanation and *theory* of some sort, because statements backed by reason usually have a persuasiveness that mere opinion lacks.

Secondly, we have to take our own team into consideration. Who are they? What can they actually do? What drove them to the pitch in the first place? In other words, we have to take *cultural* elements into consideration.

Finally, the players have to take the scheme seriously in order to give it effect. It has to carry some form of *authority* in addition to the power of the best argument. The players also have to know something about the nature of its obligations. Do they need to stick to the scheme, come rain come shine, or do they have some conditional leeway? Doctrine's three basic elements can be encapsulated in a simple figure:

In the following discussion we will take a closer look at each of the three corners, and we will also see that the concept of doctrine is often squeezed towards one of them.

Theory

A distinction can be made between 'prescriptions which belong to practical reasoning and those which logically entail an explanation of a general kind'.[18] In the case of practical reasoning we only need to know *whether* something works – not *why*. When it comes to 'explanation of a general kind' on the other hand, our belief has to be *justified*, as we ought to know why 'a postulated course of action would produce a given result before we can authoritatively recommend it to practitioners'.[19] Consequently, an

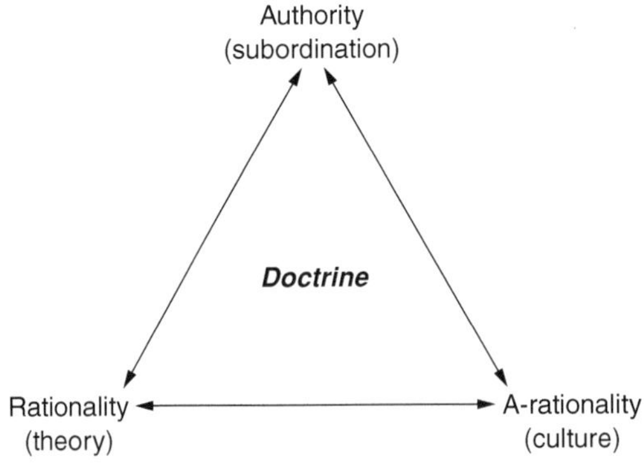

Figure 3.1 The doctrinal trinity.

important function of a doctrine will be to explain *why* the preferred doctrine is superior to all realistic alternatives. In other words, doctrine can be portrayed as 'the best available thought that can be defended by reason'.[20]

Evidently, war is a particularly hard activity to pin down in prescriptive theory.[21] War's reflexivity, complexity and contestability make a mockery out of most attempts to create an infallible recipe for victory. Consequently, exactly *what* makes an argument convincing varies across different epistemological communities.[22] The epistemological bedrock of doctrine can in other words vary across a wide spectre of supporting sciences and fields of knowledge, from mathematics to religion, from Marxism to practice theory, to name just a few.

Culture

In the other corner of the triangle you find doctrine as a cultural artefact. Culture is also notoriously difficult to define, but in this text culture is merely 'all that in human society which is socially rather than biologically transmitted',[23] and which is *a-rational.* Culture is not only an alternative to rational argument, but 'an alternative conception of reason as such, one which spurns utility and abstraction for the feel and flavour of things'.[24] For instance, I may give ample reasons why I like one type of music or one type of food, but the reasons I give are of a different order than the reasons I give to explain why a stone will sink when thrown into a lake: 'Our cultural allegiances, whether to those of our own group or to others, are not necessarily *ir*rational because they are *a*-rational.'[25]

Since little in life is more fuzzy and unmanageable than culture, it tends to cause the dream of doctrine to implode into mere dust and spilled ink:

> It is not enough to write new doctrine, if the purpose is to change the way an army will fight. Ultimately, an army's behaviour in battle will almost certainly be more a reflection of its character or culture than of the contents of its doctrine manuals. And if that culture – or mindset, if you will – is formed more by experience than by books, then those who would attempt to modify an army's behaviour need to think beyond doctrine manuals.[26]

The crucial question for doctrine makers is how deep rooted such cultural traits actually are. In order to capture the different cultural stances regarding doctrine I have drawn a model:

For culture, or a-rationality, to have any doctrinal significance at all, two preconditions have to be met. First of all culture must have casual power in one way or another. It must be able to induce people to act and cause things to happen. If not, cultural elements are mere decoration. Secondly, culture has to be manipulative in one way or another. If the makers of

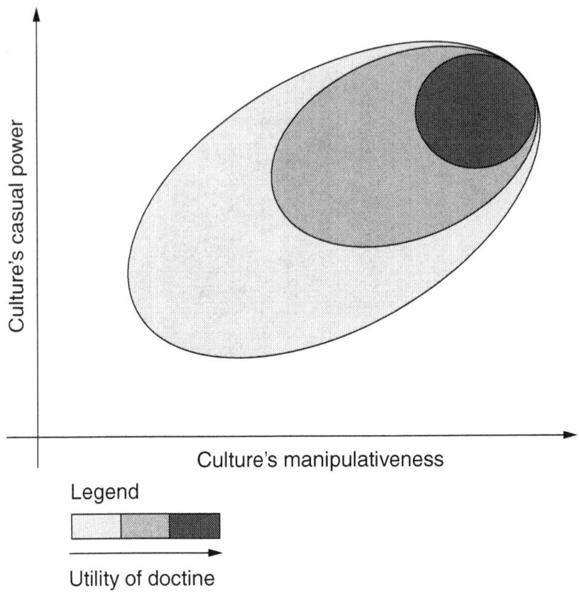

Figure 3.2 Cultural amenability.

doctrine are culture's slaves, and not its masters, doctrine is little more than a weather-vane.

Authority

Finally, in the third corner you find *authority* and authorisation. Since war is an activity where reason rarely points 'in a single direction',[27] and since military operations are an inherently collective activity, it is sometimes necessary authoritatively to choose which way to go, and to endorse a selection of theories to trust. Geoffrey Till uses an apt picture to underscore the point: 'Maritime doctrine is the application of military theory in a particular time and place. If maritime theories are about the art of cookery, doctrine is concerned with today's menus. Both are essential.'[28] Doctrine is thus not a mere description of how things are done by experts and professionals, but a prescription telling how things should be done:

> [M]ilitary thought and doctrine are not synonymous. The first is personal, the latter institutional. A military thinker may inspire admirers to emulate or implement his ideas; but this can never be a substitute for institutional acceptance. Those who listen (or read) the works of any thinker are free to reject as much or as little as they choose from the corpus of his works.[29]

In democracies it is a truism that politicians (over)rule generals, not the other way around. Flag officers have the right to raise their voice in manners compatible with their strategic culture, but as long as the orders they finally get are legal the only option they have is to comply or resign.

As a corollary, politicians also have the right to use doctrines to convey issues that are not necessarily supported by 'the best available thoughts'. Politics can pull war, and preparation for war, in a different direction than experts, or culture for that matter, would have taken it.[30] And war itself can in return influence policy in unexpected ways.

Even if all doctrines, by this text's definition, must contain elements of all three pillars of doctrine, i.e. theory, culture and authority, not all doctrines need to emphasise all three corners to the same degree. The three pillars' weight can be varied according to need, without neglecting any of them.

The utility of military doctrine

Based on how the three elements of doctrine are balanced, you can produce three ideal types of doctrine; doctrine as a tool of education, doctrine as a tool of command, and doctrine as a tool of change, or innovation we may add.

As a tool of command a doctrine tells authoritatively what to *do*, as a tool of change it tells authoritatively what to *be*, and as a tool of education it tells what we do and who we are, for the time being.

As a *longue durée* different geostrategy has influenced the weight laid on the command side of the triangle. Nations with homogenous strategic challenges over time, for instance Germany, have tended to use doctrine, or at least doctrine-like devices, as a tool of command, while nations with heterogeneous and shifting strategic environment, such as the U.K. and

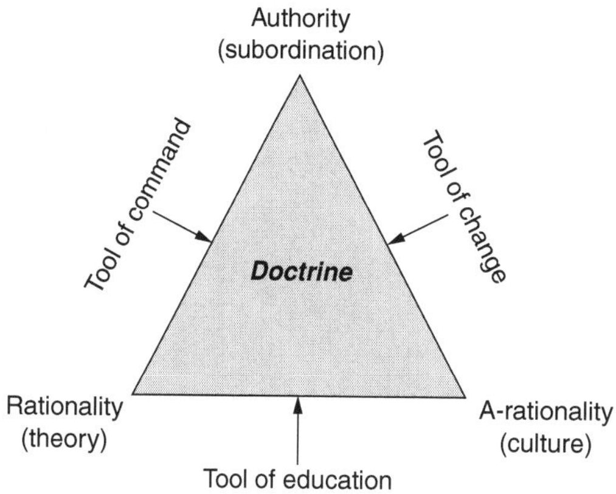

Figure 3.3 The doctrinal utility span.

the USA, have tended to use doctrine mainly as a tool of education. In addition to such deep rooted historical structures come more abrupt events, such as a shocking defeat or an astonishing victory, or new technology with revolutionary potential. Such strategic and political earthquakes may bring disaster, but they can also give golden opportunities for change. Indeed, some of the most fertile doctrinal discussions have occurred in the wake of such upheavals.

A tool of education

There is of course a plethora of military documents related to education of some sort, but doctrine is elevated from those mainly because a doctrine presents 'essential knowledge', i.e. knowledge that everyone within a given branch of military service is expected to possess. In an air force a pilot knows a lot that a mechanic does not, and vice versa, but both have to relate to a common core of basic knowledge in order to function well together.

As an alternative to providing rules and principles, a doctrine, used as a tool of education, can instead present particular cases that the military students, broadly defined, can use to reach their own conclusions. In order to do that, however, the case presented has to be detailed enough to offer sufficient similarities and dissimilarities with the situation of the reader. Military story telling is the perfect vehicle for such case presentation.

Military superstars, such as the American generals Colin Powell and Wesley Clark and the British generals Sir Rupert Smith and Sir Mike Jackson, had all retired from their military careers when they published their stories, or educational doctrines in disguise. Nonetheless, their books are important vehicles for carrying on the flame of their strong personalities, their leadership and their doctrinal convictions. In fact, it does not even take a general to produce such a *Bildungsroman*.[31] Craig Mullaney's *An Unforgiving Minute: A Soldier's Education* (2009) is a superb case in point, written by a young soldier with all the necessary credentials, such as Ranger School and West Point, in addition to a Rhodes scholarship at Oxford.

Even a formal doctrine can be electrified by linking it to a face and a history, as for instance is the case with the U.S. Army/Marine Corps *Counterinsurgency Field Manual* of 2007, which is published with a foreword by General David H. Petraeus. By adding Petraeus' name to the document it ties it into a greater story, i.e. Petraeus' personal history. In addition to being one of the doctrine's main contributors, Petraeus is a man who combines bravery in battle, wisdom in war, and a Ph.D. in international relations, perhaps a perfect combination for fuzzy conflicts. Even if doctrine can be used profitably as a tool of education, the core function of doctrine is still usually seen as a tool of command.

A tool of command

Doctrine's rationale as a tool of command is to enhance military cohesion and facilitate coordination while simultaneously reducing the need for military commanders to issue detailed instructions. In short, the aim of doctrine is to reduce the effect of Clausewitzian friction.

A simple agreement is usually the most effective way to solve coordinating problems, but in combat, where things can happen fast and involve many people, agreement may be out of reach. However, a doctrine can facilitate tacit conventions in the heat of battle by prearranging salience and distinguishing relevant precedents. If everybody in a unit instinctively comprehends the situation in a similar way and trusts everyone else to do the same, countermeasures can be taken more swiftly without cumbersome and time consuming coordination. Obviously, we are balancing on the horns of the 'basic doctrinal dilemma' here.[32] If our troops act like robots or mindless imitators, their performance would be judged outstanding at a military tattoo, but they can be pretty easy to outsmart.

Doctrinal critics have thus always questioned the wisdom of issuing authoritative documents regulating behaviour or comprehension in combat, and have instead opted for a '*doctrine of no doctrine*'.[33] To them it is fairly obvious that each situation has to be treated on its own merit. Especially in Great Britain, which has fought its wars worldwide, many have seen the multifariousness of war as the main argument against doctrine. It can 'prepare the army to face the wrong army at the wrong time and in the wrong place'.[34]

However, it is exactly *because* of this diversity of missions that you need a doctrine. If any army can use a doctrine, it is those with multifarious tasks:

> The task of training the great armies of the Continent is child's play compared with the severe and intricate problem which is imposed upon our General Staff by the geographical distribution of the units which compose the British Army.... That the task of ensuring unity of effort from men trained under such diverse conditions is insuperable unless their minds are guided and bent in some one definite direction by the assimilation of a common doctrine is recognised by the General Staff.[35]

By nature it is difficult to anticipate what our adversaries will do and even sometimes what our own politicians are likely to do, but a common doctrine makes it at least easier to guess what our own troops will do. Hence, a doctrine of no doctrine imports into the problem 'fresh unknown factors – namely the unknown minds of subordinate commanders.'[36] Thus, if you reject formal doctrine in order to make your troops inventive, all you accomplish is to invite a multitude of unknown doctrines in by the backdoor:

> Instead of a General Staff doctrine, held in common with comrades and chiefs, each commander will have his own doctrine.... In short, adherence to the 'doctrine of no doctrine' intensifies the difficulty of transferring the mind of the commander to the minds of his subordinates, a difficulty which it is the object of all training and teaching to overcome.[37]

Even in modern COIN operations, a common doctrine wisely written will not hamper the mid-ranked decision makers, but assist them. Doctrine is not a screenplay telling how you should act in all situations, no more than conventions in general are. Doctrine is not ossified habits that make you predictable and sluggish, but a springboard to bounce off. Moreover, if you think that a doctrine may lure you into preparing your army to face the wrong enemy no preparation at all is presumably not the answer.

It is strategy and operational art that translate political aims into military tasks, but without a doctrine, without a grammar so to speak, such translation becomes difficult. The grammar is not there to restrict your eloquence, but to make you understandable to your surroundings.

Strangely, there is often an implied assumption that an armed force can have only one doctrine. However, just as it is fully imaginable that a person can be both a parish priest and rugby player without ever mixing the two, or that a police squad can treat gangsters substantially different than school kids, it is at least imaginable that an armed force can do two completely different things. At the deepest organisational level there is perhaps room for only one capstone doctrine, but on a more practical level there can be several doctrines at the doctrinal tree. Having more than one doctrine is not doctrinal cheating, it is doctrinal necessity.

Perhaps new technology and the cheap access to information subvert the rationale for doctrine as a tool of command? If your boss knows a lot more than you do about the situation you are in, he would certainly not leave it to your judgement (however doctrine assisted) to sort out the situation. When the digitised battlefield enables the President virtually to monitor the heart rate of any given grunt, the temptation to strategically micromanage may become irresistible.

However, even this enticement, on all command levels, can be subject to doctrine, a doctrine not primarily for the boots on the ground, but for the executives, for those holding 'the long screwdriver'. Perhaps the most pressing doctrine in the future might be that which tells the commanders to do less?

As such, doctrine as a tool of command may make for a dramatic U-turn, in the sense that while its original *raison d'être* was to be a beacon for military commanders operating in the informational darkness, it can now become a floating device for use in the tremendous ocean of information made possible by modern information technology. Doctrine can be a pilot guiding you through the Scylla of uncertainty and the Charybdis of information overload.

A tool of change

As mentioned previously, a doctrine can act like a weathercock or a thermometer, only revealing tendencies and policies it is ultimately unable to do anything about. Some commentators claim that this is the normal situation, especially on the strategic level.[38] However, glancing at history, it seems obvious that doctrine can influence a wide spectrum of activities, such as procurement, fighting, planning and training.[39] In other words, doctrine can be an important device to transform ideas into operational hardware and new ways to operate, as mentioned in the opening of this chapter.

As a rule, doctrine as a tool of change has been most in demand in the wake of military disasters and other revolutionary changes in the strategic context, where practical and cultural changes were sought and speeded up by a controlled revolution from above. In order to substantiate that claim we will pay a short visit to some particularly volatile periods in the last 200 years of military history.

One of the greatest military eye-openers in military history occurred north of Jena in 1806. It was 20 years since Frederick the Great had died, but most Prussians still regarded the Frederickian system as superior to all others. Prussia had been at war against revolutionary France in the 1790s, but had yet to meet Napoleon in open contest on the battlefield. When that eventually happened, in 1806, it ended in utter disaster.

In Frederick's world, iron discipline was the pinnacle of military prowess. Standing firm in the face of mortal danger and firing in unison without aiming was the core of military excellence.[40] For Gerhard von Scharnhorst, on the other hand, this attitude was no longer supreme prowess, but supreme stupidity no less. Old habits had to be weeded out: 'Before the war we taught the men to load quickly, but not well, to fire quickly, but without aiming. This was very ill-considered; we must therefore work with all our might to root out this error.'[41] As argued previously, the Prussians did not use the word doctrine as we know it today, but the Prussian resurrection after 1806 shows at least that it is possible to change quite a lot relatively swiftly.

The French willingness to change, and their willingness to expose their despair after the disaster in 1870 was, as also seen previously, somewhat unprecedented. Their supreme commanders' inexplicable passivity on the battlefield, as for instance at the Battle of Gravelotte in August 1870, was turned into *élan vital* and offensive *à outrance* in the next war. Modern doctrine, as we know it today, was born in an intense desire to change and to catch up.

The American willingness to change after their bitter defeat in Vietnam was not as profound as the Prussian after 1806 or the French after 1870, but even for the Americans it was evident that something had to be done to avoid a similar failure in the future. And again the defeat could not be explained by access to material resources. One important measure the U.S. Army took was to institutionalise systematic military thinking through the establishment of *U.S. Army Training and Doctrine Command* (TRADOC) in 1973 and the issuing of a new version of FM 100–5 *Operations* in 1976, a move that triggered the doctrinal renaissance that we still are part of. Doctrine development became an 'engine of change'.[42]

Importantly however, the U.S. Army's focal point in the 1970s was not to win the next Vietnam War so to speak, but to hurry back to conventional war. Wars of the Vietnamese kind were not important enough: 'I'll be damned if I permit the United States Army, its institutions, its doctrine and its traditions to be destroyed just to win this lousy war!'[43]

Perhaps the most important thing a doctrine can do after great upheavals is to offer new *meaning*. A quest for meaning is part of human nature and can be met by different means, as for instance a new doctrine. When military employees struggle to get up in the morning and wonder why they should bother to go to work, some would find comfort in thinking about the pay cheque, or the splendour of their uniform, while others want something more. For the latter, the doctrine should be the place to look. According to Richard Dawkins, 'Religion has at one time or another been thought to fill four main roles in human life: explanation, exhortation, consolation and inspiration'.[44] This is also a rather apt description of a military doctrine, read in this particular way. Doctrine has thus kept its existential zest, especially in periods when the need for armed forces is not self-evident.

All in all, in order to significantly change huge organisations, it takes years and usually enormous efforts, in addition to commitment from powerful sponsors. Given the difficulties we should perhaps rather move away 'from trying to change organizations and instead look at how we might help them become ready for change'.[45] An important element in that process will be some sort of consciousness-raising activity, such as forging a new doctrine. If you succeed, doctrine can become 'a key weapon at the disposal of [military] leadership to steer their organisation through an era of uncertainty and sustained change'.[46] In other words, it is easy to acclaim Sir Michael Howard's doctrinal credo:

> I am tempted to declare dogmatically that whatever doctrine the Armed Forces are working on now, they have got it wrong. I am also tempted to declare that it does not matter that they have got it wrong. What does matter is their capacity to get it right quickly when the moment arrives.[47]

A fourth generation?

Military command has been through three distinct stages. The first phase was characterised by the supreme commander's presence on the battlefield.

The next stage occurred when Helmuth von Moltke definitely left the *Feldherrenhügel.* The third, and present stage, started with the end of the First World War, when airpower and increased firepower turned the battlefield into a three dimensional battle space. The introduction of new technology and the radical improvement of the old, forced military theorists to change their way of thinking, and according to Stephen Biddle the mental aftermath of the Great War is still with us.[48] Does that mean that we have seen the end of doctrinal history? Presumably not. A 'fourth generation doctrine' is budding.[49]

Military doctrines have always been issued with an eye to how it would be comprehended by adversaries and opponents, especially so with unclassified doctrines. That is part of the basic doctrinal dilemma. However, in an era characterised by wars of choice, what our own politicians, voters and taxpayers *think* the services are doing is becoming increasingly important. And new doctrines are important ways to tell them. Hence, what I call a 'fourth generation doctrine' does not have its most important readers in the armed forces.

A service may survive a lost war, but not the loss of political support. General Anthony C. Zinni of the U.S. Marine Corps was rather plainspoken on this matter:

> We teach our ensigns and second lieutenants to recognize that sister Service is the enemy. It wants our money; it wants our force structure; it wants our recruits. So we rope ourselves into a system where we fight each other for money, programs, and weapon systems. We try to out-doctrine each other, by putting pedantic little anal apertures to work in doctrine centers, trying to find ways to ace out the other services and become the dominant service in some way. These people come to me and the other CinCs and ask, 'What's more important to you – air power or ground power?' Incredible! Just think about it. My Uncle Guido is a plumber. If I went to him and asked, 'What's more important to you – a wrench or a screwdriver?' he'd think I'd lost my marbles.[50]

A particularly interesting case of fourth generation doctrine is *The U.S. Army/Marine Corps Counterinsurgency Field Manual.* Apart from the military promulgation, the doctrine was issued in 2007 by the University of Chicago Press, and can be bought in commercial bookstores. Significantly, there were journalists, human rights advocates and academics among the staff that developed it.[51] However, if doctrines are made in order to look good to modern liberal consciences, then they can send completely the wrong message to the warriors. The publisher itself indicates the problem: 'The opposing school of thought wants to save the Army from its new doctrine. Instead of worrying that the manual will not be put into practice, these critics worry that it will be.'[52] Apparently, you cannot win modern wars by being politically correct.

However, by almost mischievous ways, fourth generations doctrines may in fact turn out to be a war winning instrument exactly because they address the home front and not the soldiers in the field. That said, however, double communication can be self-defeating if there are significant discrepancies between the doctrinal message and the actual *modus operandi* in the field, especially if the soldiers start to live by the doctrine. If you must lose the war in the field in order to win it at the news desk you may be worse off than you were when you won the shootouts, but lost the arguments.

It is important not to overstate this point, particularly if it is connected directly to *The U.S. Army/Marine Corps Counterinsurgency Field Manual.* According to Stephen Biddle the *Field Manual* was in use in Iraq, but not always in a way that we would recognise.[53] Crucially, this is *not* doctrinal cheating, but military prudence: 'To the credit of General Petraeus and Ambassador Ryan Crocker, they have proven able to improvise and innovate under fire, departing from the manual to take advantage of unexpected trends'.[54] Doctrine made organisational improvisation possible.

A fourth generation doctrine is not necessarily promulgated to make the armed forces look good. It can also be issued as a warning and a clear message: 'In fact, the British Army's interim doctrine on *Wider Peacekeeping* (1995) was designed to caution policy-makers and the public about the costs of using force in peacekeeping (unusually for military doctrine, a glossy version was on sale at bookstores.)'[55]

To sum up: first generation doctrines, or rather, doctrine-like documents such as the Royal Navy's *Fighting Instructions* (1653) and Frederick the Great's *Instructions for His Generals* (1747), were made for forces where the supreme commander fought with his men, either in front of them or close behind them. In the second generation of doctrines, the supreme commanders had left the battlefield. The third generation doctrines followed the commanders and left the battlefield as well, and tried instead to orchestrate a three dimensional space and the entire range of combat capability, including physical, moral and mental domains. The fourth generation of doctrines is in danger of leaving even the war.

Doctrine – reconsidered

The question this chapter set out to answer was: What is doctrine? My short answer is that doctrine is *institutionalised beliefs about what works in war.* My elaborate answer is that doctrine is *an authoritative theory of war that allows for cultural idiosyncrasies,* where 'theory' is simply defined as:

> an attempt to bind together in a systematic fashion the knowledge that one has of some particular aspect of the world of experience. The aim is to achieve some form of understanding, where this is usually cashed out as explanatory power and predictive fertility.[56]

Despite the epistemological challenges facing all military doctrine makers, it is a modest requirement that armed forces think through their challenges and systemise their knowledge for the benefit of the future. The question is not whether armed forces ought to have a doctrine or not, but whether to be frank about it or not. Doctrines exist; it is only their degree of explicitness, status and use that vary.[57] If the doctrinal need is not filled by official doctrine it is filled by something, or someone, else. Unofficial and 'untested' doctrines will crop up elsewhere if they are suppressed at the top level:

> Doctrine should not be, and is not designed as a substitute for thought. Yet it must be said that on occasions, especially during the Second World War, an ill-designed and sometimes confused appreciation of certain ideas drawn from progressive military theory, rushed to fill the doctrinal vacuum with disastrous results.[58]

Hence, even if the development of doctrines is doomed to fail the highest expectations, the alternative is worse. We humans always theorise, in the sense that we look for patterns and connections.[59] Making a doctrine is thus a question of forethought and honesty. A doctrine cannot, presumably, make us more inventive, as argued in the opening of this chapter, but it can make us more prone to 'improvise, adapt and overcome'.

Properly written, doctrine will light your way, ease your progress, train your judgement, and help you to avoid pitfalls.[60] Alas, not even doctrine can do wonders: 'Reiterations of doctrine cannot transform human nature or "change cockroaches into butterflies".'[61]

Notes

1 Martin van Creveld, *Technology and War*, New York: The Free Press, 1991, p. 220.
2 Ibid.
3 AAP-6, *NATO Glossary of Terms and Definitions.*
4 Ladislaus Pope-Hennessy, 'The Place of Doctrine in War', *The Edinburgh Review*, January 1912, p. 21.
5 Ken Booth, *Strategy and Ethnocentrism*, New York: Holmes & Meier, 1979, p. 16.
6 Mark Mäder, *In Pursuit of Conceptual Excellence, The Evolution of British Military–Strategic Doctrine in the Post-Cold War Era, 1989–2002*, Bern: Studies in Contemporary History and Security Policy Vol. 13, 2004, p. 137.
7 Pope-Hennessy, 'The Place of Doctrine in War', p. 29.
8 'Scholars and military practitioners approach doctrine largely on the intuitive level, and there is no single approved definition. In practice, definitions follow institutional or personal whim and vary between individuals, services, nationalities and time periods.' Albert Palazzo, *From Moltke to Bin Laden, The Relevance of Doctrine in the Contemporary Military Environment*, Australian Government, Department of Defence, Land Warfare Studies Centre, 2008, p. 6.
9 Rupert Smith, *The Utility of Force, The Art of War in the Modern World*, London: Penguin, 2005, p. 96.

10 'Moltke rationalised the complexities of his age by institutionalising a self-replicant system of interpretation and action that contemporary military professionals would identify as the first articulated doctrine of war.' Palazzo, *From Moltke to Bin Laden*, p. 17.

11 'In fact, the old Prusso-German army had no concept of doctrine in the modern American and Western sense. That army used the term *Lehre* rather than *Doktrin* to describe the concepts in its official manuals.' Daniel J. Hughes, *Moltke On the Art of War, Selected Writings*, New York: Ballantine Books, 1993, p. 174.

12 The closest Clausewitz got to our concept of doctrine was presumably 'Methodismus':

> Officers whom one should not expect to have any greater understanding than regulations and experience can give them have to be helped along by routine methods tantamount to rules [Methodismus]. These will steady their judgment, and also guard them against eccentric and mistaken schemes, which are the greatest menace in a field where experience is so dearly bought.
>
> (Carl von Clausewitz, *On War*, ed. and trans. Michael Howard and Peter Paret, Princeton: Princeton University Press, 1976, p. 153)

However, as the level of war rises, method's utility 'will decrease to the point where, at the summit, it disappears completely. Consequently, it is more appropriate to tactics than to strategy' (Ibid.).

13 Robert M. Citino, *Blitzkrieg to Desert Storm, The Evolution of Operational Warfare*, University Press of Kansas, 2004, p. 34.

14 Stephen Biddle, *Military Power*, Princeton: Princeton University Press, 2004, p. 20.

15 Pope-Hennessy, 'The British Army and Modern Conceptions of War', *The Edinburgh Review* April 1911, p. 322.

16 H. Langlois, *Lessons From Two Recent Wars, The Russo-Turkish and South African Wars*, translated for the General Staff, War Office, London: HMSO, 1909, p. viii.

17 Ferdinand Foch, *The Principles of War* (1903), London: Chapman & Hall, 1920, p. 7:

> [D]octrine will extend itself to the higher side of war, owing to the free development given to your minds by a common manner of seeing, thinking, acting, by which everyone will profit according to the measure of his own gifts; it will nevertheless constitute a discipline of the mind common to you all.

18 Charles Reynolds, 'Carl von Clausewitz and strategic theory' in *British Journal of International Studies 4*, 1978, p. 184.

19 Ibid., p. 185.

20 U.S. Army Regulation 320–5, *Dictionary of Army Terms*, Washington: 1965, quoted in R. J. Spiller, 'In the Shadow of the Dragon: Doctrine and the U.S. Army after Vietnam' in *RUSI Journal*, December 1997.

21 In this chapter the word 'war' is used as it is in ordinary language rather than in a strict legal or political sense. It is not central to the main arguments whether the war is large or small, legal or illegal, etc.

22 'Epistemology' is rather broadly defined in this chapter and denotes 'the study of our right to the beliefs we have', Ted Honderich, *The Oxford Companion to Philosophy*, Oxford: Oxford University Press, 1995, p. 245.

23 John Scott and Gordon Marshall, *Oxford Dictionary of Sociology*, Oxford: Oxford University Press, 2005, p. 133.

24 Terry Eagleton, *The Idea of Culture*, Oxford: Blackwell, 2000, p. 57.

25 Ibid., p. 59.
26 Paul Johnston, 'Doctrine Is Not Enough: The Effect of Doctrine on the Behaviour of Armies', *Parameters*, Autumn 2000.
27 Eugene Garver, *Aristotle's Rhetoric, An Art of Character*, The University of Chicago Press: 1994, p. 109.
28 Geoffrey Till, *Seapower – A Guide for the Twenty-First Century*, London: Frank Cass, 2004, p. 33.
29 B. Holden Reid, *A Doctrinal Perspective 1988–98*, The Occasional No. 33, The Strategic and Combat Studies Institute, 1998, p. 13.
30 'The politicians should get what they ask for, even if it is not what they really want. In other words, the politicians have a right to be wrong.' A. Bolin, *Political–Military Relations, An introduction to a field of study*, Stockholm: Försvarshögskolan, 2004, p. 12.
31 A *Bildungsroman* is usually a portrait of a young man and:

> how he enters life in a happy state of naiveté seeking kindred souls, finds friendship and love, how he comes into conflict with the hard realities of the world, how he grows to maturity through diverse life-experiences, finds himself, and attains certainty about his purpose in the world.
> (Wilhelm Dilthey, quoted in Y. Noah Harari, *The Ultimate Experience, Battlefield Revelations and the Making of Modern War Culture, 1450–2000*, New York: Palgrave Macmillan, 2008, p. 145.)

32 J. S. McKitrick and P. W. Chiarelli, 'Defense reform: An Appraisal' in Clark *et al.* (eds.), *The Defense Reform Debate, Issues and Analysis*, p. 324:

> [M]ilitary doctrine carries a number of risks. These derive from the basic doctrinal dilemma: doctrine must be explicit and specific to achieve useful empirical content; to the degree that this occurs, however, dogmatism and doctrinal righteousness too often prevail. Efforts to avoid this pitfall often result in doctrine that is so abstract as to be of no value in the field.

33 Pope-Hennessy, 'The Place of Doctrine in War', p. 24.
34 G.F.R. Henderson, paraphrased in Hew Strachan (ed.), *Big Wars and Small Wars, The British Army and the Lessons of War in the 20th Century*, Abingdon: Routledge, 2006, p. 4.
35 Pope-Hennessy, 'The Place of Doctrine in War', p. 29.
36 Ibid., p. 27.
37 Ibid., pp. 26–7.
38 'One fact of national life seems immutable; military concepts and doctrine cannot change, substantially, political ideas and attitudes concerning national defence.' D. L. Bland, 'Canada's Officer Corps: New Times, New Ideas' *CDA Institute XVth Annual Seminar 1999*. Online. Available at: http://cda-cdai.ca/seminars/1999/99bland.htm (Accessed 25 January 2009).
39 '[C]hange of doctrine cannot be entirely ignored, for adopting a new doctrine can result in substantial changes in the practices and structure of a military organization.' T. Farrell and T. Terriff, *The Sources of Military Change, Culture, Politics, Technology*, Boulder: Lynne Rienner, 2002, p. 5.
40 'Aiming was in any case forbidden since it would have slowed the rate of fire – at most, the men were told to point their weapons waist-high.' Peter Paret, *Yorck and the Era of Prussian Reform 1807–1815*, Princeton University Press, 1966, p. 15.
41 Roger Parkinson, *Clausewitz – A Biography (1971)*, New York: Cooper Square Press, 2002, p. 124.
42 The Army Chief of Staff General Gordon R. Sullivan ordered TRADOC in the early 90s to 'develop doctrine as the "engine of change" in an effort that would

be both "process and product" '. J. L. Romjue, *American Army Doctrine for the Post-Cold War*, Fort Monroe: Military History Office, 1997, p. 3.

43 Gunter Lewy, *America in Vietnam*, New York: Oxford University Press, 1978, p. 138.

44 Richard Dawkins, *The God Delusion*, London: Bantam Press, 2006, p. 347.

45 R. Seel, quoted in Linda Holbeche, *Understanding Change, Theory, Implementation and Success*, Oxford: Elsevier, 2006, p. 195.

46 Mäder, *In Pursuit of Conceptual Excellence*, p. 293.

47 Michael Howard, 'Military Science in the Age of Peace', quoted in Allan D. English, *Understanding Military Culture, A Canadian Perspective*, Montreal & Kingston: McGill-Queen's University Press, 2004, p. 149.

48 Biddle, *Military Power*, p. 29:

> [The First World War] introduced the central problem of modern warfare: how to conduct meaningful military operations in the face of radical firepower. And by the end of the war, an answer appeared that has remained central to great power military doctrines through more than eighty years of subsequent warfare.

49 The fourth generation doctrine is not related to William S. Lind's 'four generation warfare', apart from the fact that both recognise blurring lines between war and politics.

50 General Anthony C. Zinni, U.S. Marine Corps retirement speech. Online. Available at: www.jerrypournelle.com/reports/jerryp/empire.html (Accessed 19 May 2009).

51 J. A. Nagl, 'Foreword to the University of Chicago Press Edition', p. xvi.

52 S. Sewall, 'Introduction to the University of Chicago Press Edition' in *The U.S. Army/Marine Corps Counterinsurgency Field Manual*, U.S. Army Field Manual No. 3–24 Marine Corps Warfighting Publication No. 3–33.5, p. xxxvi.

53 'In fact, Iraq is among the cases that fit the manual's assumptions poorly – which has led actual U.S. strategy in Iraq to diverge from the manual's prescriptions in ways that are not always fully appreciated in the public debate.' S. Biddle, 'The New U.S. Army/Marine Corps Counterinsurgency Field Manual as Political Science and political Praxis' in *Perspectives on Politics*, June 2008, 6: 2, p. 348.

54 Ibid., p. 350.

55 T. Farrell, 'Humanitarian Intervention and Peace Operations' in Baylis *et al.*, *Strategy In the Contemporary World* 2nd edn., Oxford University Press, 2007, p. 315. According to Mäder, *Wider Peacekeeping's* main function was to serve as a 'generator of debate'. Mäder, *In Pursuit of Conceptual Excellence*, p. 151.

56 Honderich, *The Oxford Companion to Philosophy*, p. 870.

57 McInnes and Stone, in B. Holden Reid, *A Doctrinal Perspective 1988–98*, p. 6.

58 Holden Reid, *A Doctrinal Perspective 1988–98*, p. 13.

59 '*not theorizing* is an act [while] theorizing can correspond to the absence of willed activity, the 'default' option.' N. N. Taleb, *The Black Swan, The Impact of the Highly Improbable*, New York: Random House, p. 2007, p. 64.

60 The phrase is Clausewitz's. 'Theory then becomes a guide to anyone who wants to learn about war from books; it will light his way, ease his progress, train his judgment, and help him to avoid pitfalls.' Clausewitz, *On War*, p. 141.

61 S. L. A. Marshall quoted in Jay Luvaas, 'Some Vagrant Thoughts on Doctrine' in *Military Review*, March 1986, p. 60.

4 The impact of the Office of Net Assessment on the American military in the matter of the revolution in military affairs

Stephen Peter Rosen

Introduction[1]

What was the nature of the role of Andrew Marshall and those who worked for him in the matter of what is called the Revolution in Military Affairs, or the impact of information technologies, broadly defined, on military concepts of operation? This question is an important one in the context of this book, for at least two reasons.

First, the conventional account is intellectually incomplete. The conventional account, briefly put, is that Mr. Marshall read Soviet military journals that referred to an ongoing Military Technical Revolution, associated with the use of sensors and guided munitions that would give the United States the ability to neutralize Soviet tank formations with non-nuclear weapons on a geographic and temporal scale that had previously been achievable only with enhanced radiation nuclear weapons. He circulated the idea that such a revolution was taking place in an assessment written by Andrew F. Krepinevich in 1992.[2] In the wake of the unexpectedly rapid and decisive victory of American airpower over Iraqi ground forces in 1991, this assessment attracted the attention of the senior leadership of the American military, and generated a wave of interest in military 'transformation' that reached its peak in 2001, but which was then submerged by the need to fight wars of counter-insurgency in Afghanistan and Iraq in the aftermath of the September 11 attacks on the United States.

This account is incomplete because it does not answer the question of why Mr. Marshall paid attention to Soviet doctrinal writings in the first place. The possible significance of these writings was noticed by few, if any, other Americans. Moreover, his receptivity to the argument in Soviet journals that a comprehensive revolution in military concepts was under way was at odds with the general tendency of Americans to take a pragmatic and task oriented approach to technology. Dima Adamsky has characterized the American technological style well. The United States, he argues, has an intellectual culture that is empirical and problem oriented and less inclined to see the systemic character and impact of incremental improvements in technology. In addition, the United States lacked the kind of General Staff that focused on broad questions of the changing character of war, in the manner of the pre-World War II German Staff or the Soviet General Staff.[3] Yet Mr. Marshall was not only receptive to the Soviet perspective, but urged people in conversation to think about the possible systemic impact of this revolution on trans-oceanic power projection, an issue that, to my knowledge, was not raised at all by the Soviet writings. Mr. Marshall's thinking, thus, was broader and more comprehensive than the Soviet writings, yet it was distinctly un-American. Where did this perspective come from?

Second, the account is incomplete because it does not adequately capture the nature of the American reaction to Mr. Marshall's ideas. A closer examination of the historical record indicates that it was not simply the near term pressures that followed from September 11, but deeper patterns of organizational preferences.

In the following sections of the chapter I will briefly review the dominant American perspective on the technological changes we associate with the Revolution in Military Affairs in the period beginning in the 1960s and leading up to the 1990s. This was the context in which Mr. Marshall advanced his own views at the end of the 1980s and the beginning of the 1990s. I will then set out what appears to me to have been the particular perspective brought to this matter by Andrew Marshall. Third, and finally, I will sketch out what the Office of Net Assessment did to try to promote thinking by the American military on this subject, and what the American military did in response.

Dominant American perspectives on RMA technologies

Adamsky argues that the tendency of nations like the United States, and military organizations without a Soviet style General Staff, will be to understand technological change in a disaggregated, case by case manner. Possible improvements in the capabilities of an individual weapons system will be seen as only that, valuable, but to be judged in terms of whether they deliver incremental improvements in the ability to execute existing missions. Other scholars of American military technology confirm this assessment. Barry Watts, for example, has documented the early introduction of guided munitions into the American Air Force and Navy beginning in World War II. His meticulous study shows how the early use of radar guided air-to-air missiles (the SPARROW AIM-7) in the Vietnam War was understood in narrow, task oriented terms. Specifically, the revolutionary potential of these weapons was underestimated because of the unreliability of vacuum tubes and hand-soldered connections in the first generation of weapons that were deployed in combat. The substitution of solid-state electronics into the same systems later drove up their reliability and their effectiveness, but the early problems reduced their impact on Air Force thinking.

Similarly, the early use of very accurate air-to-ground weapons in the Vietnam War was not recognized as potentially revolutionary. In the first use of laser-guided bombs in that war in 1969, the United States Air Force released 1,601 2,000lb bombs with laser guidance packages, of which 85 percent hit within 9.6ft of their targets, with an operational reliability rate of 85 percent. In the Linebacker air campaigns against North Vietnam in the period February 1972–February 1973, more than 10,500 laser-guided bombs were released on targets and 9,107 landed within 25ft of their targets. This compared with an average circular error probable 500ft of F-4's dive-bombing heavily defended targets. The Air Force histories of this campaign, however, neglected the new weapons and emphasized the impact of B-52 strikes using unguided bombs.[4] Later accounts of the first Gulf War also provide material which suggest why the new guided air-to-ground munitions were not appreciated by the Air Force. The United States commander of the air forces deployed Lt. General Charles Horner to the CENTCOM Theatre in 1990. He was briefed by Colonel John Warden on August 20 on the plan for the bombing campaign against Iraq, Instant Thunder, that utilized precision-guided munitions to execute rapid accurate strikes against Iraqi command and control targets at the start of the war. Horner is reported to have reacted as follows:

> No sooner had Warden begun the briefing than Horner began to assault him with criticism, complaints, and questions, deriding the advice he was getting.... Warden's discussion of precision-guided munitions was off base. Warden should be talking about the precision-guided delivery of weapons. This was a reference to the F-16, one of Horner's favorite weapons, which did not have the capability to deliver laser-guided bombs, but which had the supposed capability to drop its dumb bombs with great accuracy at low altitude using its radar bombing system.[5]

To be sure, there were genuine limits on the utility of the first generation of laser-guided weapons. They were useable only in good weather, and they were still relatively expensive. But perhaps more importantly, laser-guided munitions reduced the importance of fighter pilot skill in the delivery of weapons, and so were a challenge to the importance an established Air Force social hierarchy based on pilot skill. There was, as a result, powerful social pressure to see these weapons not as something that would require a major change in military concepts of operations, but as simply one more way of doing what the fighter pilots of the Air Force were already doing, as, in other words, an incremental improvement.

In addition to precision-guided weapons, the ability to deliver information about the location of moving targets fast enough to direct weapons was crucial to the Soviet concept of the Military Technical Revolution. But from the standpoint of the United States, tactical command, communications and control had been something the United States Army had for decades been trying to improve, with great difficulty and with only limited results. In her thorough history of Army tactical command and control up to the 1980s, Elizabeth Stanley saw only slow and marginal progress, which had not changed the core tasks of the Army.[6] The judgment that Army command, communications, and control was only improving incrementally was generally shared at MITRE, the Federally Funded Research and Development Center (FFRDC) devoted to that mission area.

While far from being a comprehensive survey, these observations tend to support the view of Adamsky that the United States would not, and did not, understand the revolutionary potential of the developments that had been identified by the Soviet military. But Mr. Marshall did. Why?

Intellectual background of Andrew Marshall and the RMA

While the Soviet MTR and the American discussion of the RMA focused on tactical and operational non-nuclear war-fighting, the work done by Mr. Marshall at the RAND Corporation in the period 1950–1969 revolved around the problems of intercontinental nuclear warfare, and appears to have prepared him to understand the impact of information technologies on the conduct of warfare as revolutionary, not incremental, system-level improvements. The centrality of deterrence in American thinking about U.S.–Soviet strategic nuclear interactions led him to be concerned by how it was that the Soviet leadership thought about the character of a possible exchange of nuclear weapons. Deterrence existed in the mind of the adversary, and so understanding the mind of the adversary was crucial to the task of designing and building an adequate deterrent capability. Beyond that, critical events led Mr. Marshall to see, concretely, that American efforts to think like Soviet leaders were seriously flawed, because they failed to take into account the political and organizational imperatives that shaped Soviet thinking, and which led the Soviet military in directions that differed significantly from the behavior that Americans calculated to be rationally optimal. All this prepared him to take seriously the judgment of Soviet thinkers about the revolutionary nature of the change in warfare produced by information technologies.

Specifically, in the days before the availability of good overhead reconnaissance intelligence about the Soviet Union, RAND analysts put themselves into the shoes of Soviet military commanders to guess what was going on inside the Soviet Union. The inability of Americans to think like Soviets led them into gross error. It was assumed, for example, that the Soviet Union would build its bomber bases far from the coasts and along the trans-Siberian railroad lines to make them less vulnerable to American strike forces. This is what the RAND analysts would have done if they were picking the sites of Soviet bomber bases. In fact, Soviet bases were located in the eastern areas of the Soviet Union and on the Far Eastern coast, not in the interior of Russia, because that was where Soviet air bases had been constructed in the 1930s and 1940s. Mr. Marshall, along with Joe Loftus, who had come to RAND from Air Force Intelligence, was one of the few analysts at RAND who had access to sensitive signals intelligence that showed that this assumption about Soviet air bases was completely wrong.

It was similarly assumed that because the Soviet Union focused on land war in Europe, the Soviet air forces would be dominated by tank-buster aircraft. But the Soviet experience of a decisive German air strike against Russian airfields at the beginning of World War II and the prospect of American nuclear attack had led the Soviet Air Force to devote the bulk of its resources in this period to air defenses, not ground attack aircraft. It was assumed that in the 1960s and 1970s the Soviet military would assess the hardness of American Minuteman ICBM missile silos the same way that the United States did, and that the Soviet military would understand that they would survive a Soviet ICBM attack, and were thus designed to be part of an American retaliatory force. In fact the Soviet military assessed the silos as being far more vulnerable than did the United States, and so assumed that the Minuteman missiles could only be part of an American first strike.

Discrepancies such as these led Mr. Marshall to take seriously what the Soviet military wrote, rather than assuming that they would do what Americans would do. What the Soviet doctrinal writings in the 1950s were writing about were military revolutions. They referred to the introduction of nuclear weapons, long range missiles and 'cybernetics' or the first generation of data processing capabilities, as a revolution, one of a series of revolutions that included the revolution created by the mechanization of ground warfare combined with aviation combined with poison gas used for military purposes, known in the West as *blitzkrieg*. In addition, Mr. Marshall's unusual access to intelligence about the Soviet Union's nuclear forces gave him a sense of the scale of the transformation the Soviet military had undergone to deal with the nuclear revolution, in terms of its force structure, and concepts of operation.

Independently, Albert Wohlstetter also was focused on problems of intercontinental strategic warfare. In his case, however, he was focused on making deterrence more credible and more useful by reducing the large scale damage to civilian populations that would result from strategic nuclear attacks on Soviet military targets. Wohlstetter developed what he called the 'dual criteria', or being able to destroy targets in intercontinental warfare while not doing unwanted collateral damage. This led Wohlstetter to be interested in improving the accuracy of strategic nuclear weapons. This would make it possible to reduce their yields without reducing the amount of intentionally inflicted damage that they could do. Wohlstetter came to direct a study carried out for the Defense Advanced

Research Project Agency in 1975, a study that involved a group of analysts from the military and defense industry including Mr. Marshall. It produced what was titled the *Long Range Research and Development Planning Program*. One of the major findings of the report was that a revolutionary change in the character of intercontinental strategic warfare was imminent. Improvements in accuracy would make it possible to give strategic weapons armed with non-nuclear warheads the destructive power against military targets that previously had only been achievable with nuclear warheads. To quote the report, the observable trend:

> very strongly suggests that non-nuclear weapons with near zero miss may be technically feasible and militarily effective. If so, such non-nuclear weapons, under a wide range of circumstances, might satisfy the United States and Allied damage requirements that now require the use of nuclear weapons. Near zero non-nuclear weapons could provide the National Command Authority with a variety of strategic response options as alternatives to massive nuclear destruction.[7]

There is no reason to believe that Mr. Marshall shared Wohlstetter's view that strategic level deterrence would be enhanced by reducing the collateral damage done by nuclear weapons. Indeed his written work at RAND in 1960 argued that deterrence of the Soviet Union could be enhanced by the ability to inflict very large scale damage on Soviet urban centers. Very large yield nuclear weapons would reliably destroy both Soviet military targets and large portions of the urban/industrial centers that played a major role in Soviet war production and post-nuclear war recovery plans. Nonetheless Mr. Marshall was in contact with Albert Wohlstetter in this period, exchanging ideas with him formally when his office was directed to support the newly established Defense Policy Board, created in 1982 and chaired by Ambassador Seymour ('Sy') Weiss and on which Albert Wohlstetter and Mr. Marshall both sat. At the meetings of the Defense Policy Board in 1982 and 1983, Wohlstetter made the case not only for much more accurate strategic weapons (long range cruise missiles employing radar or optical terrain-mapping techniques to update their location information were achieving greater accuracies, but could be made even more accurate with then available technologies), but for a new intelligence database to support their operations in wartime. Very accurate low yield non-nuclear strategic weapons would require information not about the location of factories or military bases but about the floor plans of enemy buildings and offices so that target planners could know which room, or even which corner of a room, to hit with a weapon so that the right piece of machinery would be destroyed without destroying the rest of the building. This may have been one of the first occasions on which major shifts in American force structures, related command and control, and intelligence functions were proposed on the basis of emerging technologies, in a manner that was recognized as revolutionary, because it would substitute non-nuclear for nuclear weapons.[8] It was a clear precursor of the Presidential Commission on Integrated Long Term Strategy chaired by Wohlstetter and Fred Ikle and which produced a final report entitled 'Discriminate Deterrence'.[9]

The long term defense competition between the U.S. and the Soviet Union

One of the major intellectual contributions made by Mr. Marshall in the 1970s was the development of the concept of long term peaceful competitions between opposing defense organizations. He argued that the persistent sensitivities and predispositions of organizations could be triggered by actions taken by their adversaries as they developed their own forces. Some of those reactions were more harmful than others. By understanding those sensitivities and predispositions an adversary could, over a period of time measured in years or decades, choose what to build and when to build it in order to elicit the least harmful of the possible enemy reactions. By understanding the fears and sensitivities of an adversary, programs could be initiated or reinforced in ways that reduced the confidence of the adversary in his ability to win an engagement or a war. This could enhance deterrence.

As was the case with the study of strategic nuclear deterrence, the focus on the long term competition with the Soviet Union led Mr. Marshall to become very interested in Soviet perceptions and sensitivities. In the period beginning in 1978, the Soviet General Staff began to take notice of new United States programs for the NATO area that were meant to redress the emerging quantitative and possibly qualitative superiority of the Soviet forces facing NATO. These included thinking by then USAFE commander General John Vogt, the officer who had pioneered the use of laser-guided weapons against North Vietnam in 1972, on very deep air strikes into Warsaw Pact territory to disrupt the echeloned forces en route to the battle front with NATO. It also included a program that came to be called 'Assault Breaker' that used airborne sensor platforms to conduct wide area searches to locate armored formations on the move, and direct ground-based missile launchers to fire missiles towards those formations. When over those formations, they would dispense autonomously guided submunitions that could home in on and destroy tanks on the move. This weapons system was characterized by the Soviet General Staff as a reconnaissance-strike complex, and was assessed as having the destructive capacity against a tank battalion equivalent to a one kiloton enhanced radiation warhead.

Initially this was not perceived by Mr. Marshall as a revolution in military affairs, but as a Soviet sensitivity or vulnerability that could be exploited in the long term competition with the Soviet Union. It would enhance deterrence or force the Soviet Union to divert resources away from offensive capabilities and towards defenses against this new problem, to include, for example, long range air defense missiles that could target the airborne wide area sensors. Soviet military leaders were very sensitive to American developments in this area, in large part because they were concerned that they did not have the necessary competence in electronic data processing to compete with the United States in this area. This thinking became public in May of 1984 with the publication of an article in *Red Star* by Marshall Ogarkov assessing the impact of wide area sensors, guided munitions and automated command and control.[10]

But in addition to the Soviet reaction to American technological developments, there is circumstantial evidence that there was very sensitive intelligence available to a small number of Americans that sensitized them to the possibility of a military revolution being developed by the Soviet General Staff. This intelligence may have increased the general awareness of the possibility of ongoing military revolutions. Specifically, it is possible that Mr. Marshall was affected by the availability of the human intelligence coming from the Polish officer, Col. Rsyard Kuklinski (who was assigned to the Soviet General Staff), and the Soviet officer General Dimitri Polyakov (who was assigned to Soviet military intelligence headquarters in Moscow in the period 1976 to 1979). These two officers appear to have had access to General Staff documents outlining the Strategy of Deep Operations (Global and Theater), which had been compiled in the period 1977 to 1986. This information was circulated within very limited circles in 1981. The American defense intelligence officers who managed and analyzed this material included John Hines, who came to work for Mr. Marshall, and Phillip Petersen. The work of these analysts suggested that the Soviet Union sought to wage and win a war on a continental scale within a very brief period of time, without the use of nuclear weapons but using instead newly formed Operational Maneuver Groups that integrated ground, fighter bombers, special operations, and engineering units. This doctrinal work was associated with Soviet military exercises in the late 1970s, which culminated in a massive military exercise in September 1981 – *Zapad*-81. In the words of Diego Ruiz-Palmer, who worked for Mr. Marshall at that time, 'we monitored this exercise intently. [Marshal] Ogarkov showed that over the past three years he had studied the ways in which the Americans' "assault-breaker" would work, and what actions could be taken to minimize their impact.'[11]

This intelligence may have stimulated the thinking of those who were exposed to it. It made them very conscious of the revolutionary possibility of warfare extended in geographic scope, compressed in time, and succeeding without the use of nuclear weapons on the battlefield. In other words, the Americans exposed to this intelligence were not asked to believe in a possible American revolution in military affairs, but observed an actual revolution in military affairs that was being developed and exercised by the Soviet Union.

The intelligence work may have continued through United States Army channels. General William Odom, then Army Chief of Staff for Intelligence, and with whom Mr. Marshall had contact in the early 1980s, sponsored his own work to analyze and codify Soviet military doctrine in what he called 'the Soviet Battlefield Development Plan' in order to guide the development of American military doctrine to counter it.[12] Mr. Marshall did have contact with General Odom in the development of the assessments of the military balance in the NATO–Warsaw Pact theatre, and in the East Asian theatre. One of the members of Odom's staff, Major Garrett Ruehl, was recruited to work for Mr. Marshall in 1982 or 1983 on the East Asian military balance. Odom then published an article in *Foreign Affairs* in the winter of 1988–1989 in which he set out Soviet thinking on historical revolutions in military operations.

In summary, we can retrospectively observe a stream of military intelligence, beginning in 1981, on Soviet thinking about an ongoing revolution in military affairs that would make possible the use of non-nuclear weapons that would have an impact on ground warfare comparable to that which could only have been achieved with tactical nuclear weapons. Mr. Marshall was exposed to this intelligence and was primed to ask the question that led to the 1992 assessment: was the Soviet General Staff right to assert the existence of a revolution? His attitude was certainly the result of his ongoing search for points of sensitivity and vulnerability in the Soviet military, and may have also reflected his early exposure to sensitive intelligence and the thinking of Albert Wohlstetter.

The impact of Mr. Marshall on the American military

At least two sets of events were set in motion in the years 1990 and 1991. First, after Mr. Marshall's 1990 request to Andrew Krepinevich to complete an assessment of Soviet writings on the military–technical revolution (but before that assessment was released) the Office of Net Assessment sponsored a series of small discussions/seminars involving American military officers. Word of these seminars led General Gordon Sullivan, in the summer of 1991 while he was still Vice Chief of Staff of the United States Army, to approach Mr. Marshall with an offer to become his 'partner' in sponsoring work to explore how the Army should think about the future of war. Similarly, the Chief of Staff of the United States Air Force, General Merrill 'Tony' McPeak called on Mr. Marshall in late 1991 or early 1992 saying that he personally wanted to be the Air Force 'point of contact' on matters related to Marshall's work on the military technical revolution. Mr. Marshall's ideas had clearly caught the attention of the leadership of those two services.

Second, the first Gulf War took place in the first months 1991 and the Soviet Union collapsed. Although the first Gulf War is properly understood as perhaps the last mass industrial war fought by the United States, it did include elements that featured what might become elements of future modes of warfarc. Whilc thcrc had been arguments about the efficacy of airpower against ground targets, Barry Watts, in his analysis for the Gulf War Air Power Survey, unambiguously highlighted how the character of air war was changing. He compared missions flown by F-111 bombers against ground targets in Iraq that dropped unguided munitions, and missions against comparable targets flown by F-111 bombers with laser designators dropping laser-guided munitions, and established that that latter were able to destroy 13 times as many targets per airplane.

It is likely that the latter set of events led to the first set of events. Those events also probably helped Mr. Marshall in his effort to encourage the services to think about the implications of the military–technical revolution for their future operations. Mr. Marshall never stated what he thought the shape of future American military force structures or their concepts of operation should be. He regarded this, intellectually and morally, as a decision that the services had to make for themselves. He did, however, wish to facilitate service efforts to think this issue through, and to that end sponsored a series of seminars with the services. These began in Annapolis in 1992 when the senior leadership of all four military services assembled alongside senor figures from major American corporations who had experienced major transformations. These included senior figures from AT&T who had experienced the break up of their monopoly into competitive 'Baby Bells'. The discussions were meant to help the American military think through how large organizations changed the core nature of their operations. This seminar was followed by other seminars with individual services.

Although the Gulf War encouraged the services to take Mr. Marshall's ideas seriously, the war itself combined with the collapse of the Soviet Union seems to have led the services to the conclusion that while there might be a revolution in progress, they were *already* engaged in transforming themselves, and therefore did not need to make any significant changes. At the 1992 Annapolis Conference, the observer from AT&T listened to the American military leadership discuss what they planned to do and then commented that they sounded like the AT&T management before the break-up, when they did not realize the magnitude of the changes that were necessary. These managers, he commented, all had to be fired.

At a conference on the RMA with the Air Force, one participant listened to the Air Force's plans for technology acquisition and commented that though these were impressive plans it sounded like the Air Force was staying in the same business: compiling lists of targets, and then dropping bombs on them. This, it was commented, was not a revolution. In response, one Air Force General commented, 'you are right'.

Discussions with the United States Marine Corps were initially more encouraging, because the Marines undertook a serious program of experimentation to explore new concepts of operations. This was in itself a major change, since real experimentation was foreign to a military that tended to use the initial utilization of equipment or concepts not as part of an effort to gain knowledge, but as part of a program to validate decisions that had already been made. Marine Corps experimentation did lead to a new concept of operations, called 'Sea Dragon', that abandoned the core mission of the Marines – amphibious assault – in favor of operations that proceeded directly to the target from ships or bases far from the coastline. But this initial effort appears to have encountered serious opposition from within the Marine Corps. On one occasion a briefing on 'Sea Dragon' elicited a critical comment from a Marine Corps general officer, who commented 'when I asked you to think out of the box, I expected you to stay within sight of the box'. The United States Navy also began a strong effort to think about new concepts of operations, but that effort still does not appear to have produced major transformations in the major sub-communities within the Navy.

The games that were sponsored by Mr. Marshall also appear to indicate that his interest in fundamentally new modes of warfare did not extend to offices outside his own. Consider the differences in the war games sponsored by the Office of Net Assessment and the war games sponsored by other offices. One survey of 'futures' war games completed in 1998 showed that the ten games sponsored by Mr. Marshall in the period 1994–1997 featured large, technically capable adversaries, who attempted to use new capabilities to thwart older American power projection forces, and attacked the sources of information used by the United States to block American military objectives, while the American players investigated new ways of projecting power, from undersea and space, for example. These were games exploring a possible revolution in military affairs. The games sponsored by other agencies investigated how to integrate reserve forces into American military operations, achieve mobility on the battlefield and use terrain, the importance of sociology for special force operations, urban warfare, and new systems such as rail guns. While space-based operations were discussed, new missions and concepts of operation received less attention.[13] These latter games explored, at best, how to conduct existing missions more effectively.

If this assessment is correct it could plausibly be argued that the Gulf War both encouraged the services to discuss the Marshall's ideas, but also deprived them of a motive to change themselves in non-incremental ways. The dissolution of the Soviet Union and reductions in American military spending probably contributed to a tendency to defend existing missions, while the second Gulf War and the war in Afghanistan led to an urgent focus on the problems related specifically to those wars and to the problems of counter-insurgency: all of which took time and money away from the exploration of revolutionary change.

Notes

1 This essay was written on the basis of my reading of generally available material and on my personal impressions, and does not represent or reflect the views of Andrew Marshall, the Office of Net Assessment, or the Office of the Secretary of Defense. Because it does not rely on the internal documents of the Office of Net Assessment or any component of the United States Department of Defense, and because it reflects my own personal judgments, and in some cases, speculations, it may contain errors or omissions, though I have done my best to avoid both.

2 Andrew F. Krepinevich, *The Military Technical Revolution: A Preliminary Assessment,* Washington: CSBA, 2002. Online. Available at: www.csbaonline.org/4Publications /Archive/R.20021002/R.20021002.MTR.pdf.

3 Dima Adamsky, *The Culture of Military Innovation: Comparing the Revolution in Military Affairs in Russia, the U.S. and Israel,* Stanford CA.: Stanford UP, 2010.

4 Barry D. Watts, *Six Decades of Guided Munitions and Battle Networks: Progress and Prospects,* Washington D.C.: CSBA, 2007, pp. 184, 190, 9.

5 Michael R. Gordon and Bernard E. Trainor, *The Generals' War,* Boston: Little Brown, 1995, p. 92.

6 Elizabeth A. Stanley, *Evolutionary Technology in the Current Revolution in Military Affairs: The Army Tactical Command and Control System,* Carlisle, PA: Strategic Studies Institute, Army War College, 1998.

7 *Long Range Research and Development Planning Program,* Washington D.C.: Defence Advanced Research Project Agency, 1975.

8 This thinking also affected my own work. See Stephen Peter Rosen, 'Conventional Combat and the Nuclear Balance' in *Journal of Strategic Studies,* 10: 1, Spring 1987, pp. 36–61.

9 Fred Ikle and Albert Wohlstetter, *Discriminate Deterrence,* Washington D.C.: Government Printing Office, 1988. Online. Available at: www.dtic.mil/cgibin/GetTRDoc?Location=U2&doc=GetTRDoc.pdf&AD=ADA277478.

10 Referenced by Mary C. Fitzgerald, 'Marshall Ogarkov on the Modern Theater Operation' in *Conflict Quarterly,* Summer 1986, pp. 39–58.

11 Gordon S. Barrass, *The Great Cold War,* Stanford, CA: Stanford University Press, 2009, pp. 211–16.

12 The Odom papers on this subject and others have been donated to the United States National Archives.

13 William J. Hurley *et al.*, 'Summaries of Recent Futures War Games' in *Joint Advanced Warfighting Program,* Institute for Defense Analysis, Alexandria, VA, October 1998.

5 Restoring the primacy of battle

U.S. military theory and the RMA

Antulio J. Echevarria II

The decade prior to September 11, 2001, was a rich one for U.S. military theory. However, this abundance was not limited to good ideas.[1] Contrary to popular opinion, American military tradition – shaped as it is by military professionals and civilian defence intellectuals – has never been shy about theorizing. This fact runs counter to the received view that Americans as a whole disdain theory, and prefer practice.[2] From Alfred Thayer Mahan to Herman Kahn, American military theorists have not only been plentiful, they have been central, influencing the course of Western strategic thinking, albeit both in positive and negative ways. The sheer abundance and variety of U.S. military theory was acutely evident in the 1990s, from the revalidation of manoeuvre warfare concepts in Desert Storm to the introduction of venturous speculations in the form of network-centric warfare and effects-based operations, among others.

It was during this era, in fact, that several long-standing assumptions were called into question, such as the nature of war, what it is, whether it is changing, or even can change, and what the so-called 'new' wars of the twenty-first century portended for the roles and missions of the services. New concepts and terms, such as nonlinearity and chaos and complexity, were borrowed from other fields, while several familiar terms, such as asymmetric warfare and system of systems, seemed to gain greater currency almost overnight. By the end of the decade, U.S. military theory was in full bloom, developing rapidly and often in unanticipated directions. The events of 9/11 merely accelerated this growth, lending credence to the rather narrow and reflexive claim that entirely new ways of thinking were needed. It was not until five or six years later, in the midst of a thorny insurgency, that some of these branches could be pruned with relative certainty. By then the U.S. military was already fast developing a second grammar of war (which had to do with countering insurgencies) to add to the one it had mastered long ago, but continued to perfect, pertaining to decisive battle. The 1990s were, in fact, about using newly emerging information technologies to restore the primacy of battle.

The Gulf War: revalidating battle doctrine

As with any war in which one side completely dominates the other, Operation Desert Storm was heralded as proof of a revolutionary change in warfare. Some analysts regarded it as the forerunner of wars to come, claiming, in particular, that information-based technologies would offer the key to success on the battlefield; others saw it as a revolution in the classic sense, that is coming full circle to revalidate the time-tested, if not timeless, principles of fire and movement; still others argued that it was an anachronism, differing little in concept from the classic armoured envelopments of the Second World War. Indeed, as so often is the case, opinions ran the gamut from well reasoned to patently uninformed. The campaign was a monumental effort by any standard. More than one million troops, 10,000 tanks, 2,000 aircraft, and 100 warships arrayed on both sides.[3] For coalition forces, it was both combined and joint in character, with discreet air and ground phases, as well as a crucial maritime component. More importantly, it also gave greater visibility to the capabilities of precision-guided munitions and stealth aircraft, though these were used in relatively small numbers compared to the attention they received from the media and defence analysts.

Desert Storm validated the principles of the Air–Land Battle doctrine, as articulated in the 1982 and 1986 versions of the U.S. Army's operations manual, FM 100–5.[4] In doing so it revalidated the primacy of battle in U.S. military theory, as evidenced in the idea of identifying 'the enemy's operational center of gravity and the concentration of superior combat power against that point to achieve decisive success'.[5] Ground commanders attributed victory to the synchronization of mobility and firepower throughout the depth of the battle area.[6] Yet, for all the evidence of an integrated battle, of simultaneous attack in depth, the operation did unfold in more or less traditional phases: (1) establish air superiority, (2) knock out enemy armour and artillery pieces, and (3) launch the decisive ground assault. Put differently, the actual execution of the campaign was not entirely reflective of doctrinal rhetoric.

For the U.S. military, Desert Storm represented a break with the restrictions put in place by Cold War theory and doctrine. While Clausewitz emphasized that war had its own grammar but not its own logic, there was a sense among many American military professionals that the importance of the grammar had been lost, or diminished in some way, by the thrust of Cold War thinking, which sought to limit escalation and redefine military victory in unfamiliar ways. Some of the theories of Robert Osgood, Thomas Schelling, Bernard Brodie, and Herman Kahn appeared to ignore the Trinitarian elements of chance and hostility, and to take for granted how difficult it can be to control the dynamics of war's violence. These theorists appeared to believe as if the chain of actions and reactions in war was essentially predictable.[7] The emphasis on battle restored the importance of war's grammar, which even if it should not necessarily dictate the logic of conflict, was at least held to be inviolable. The grammar of war was considered to be derived directly, and of necessity, from war's nature, which was seen as chaotic and prone to escalation. Much of Cold War strategic thought considered large-scale combat operations as something of a last resort, particularly as the chief concern was to avoid escalation to nuclear weapons. For some military professionals, however, taking escalation off the table from the start ignored war's basic nature, and put oneself at risk by giving one's adversary the upper hand.[8]

Restoring the primacy of battle also meant settling upon one particular grammar of war. The corpus of theory concerning counterinsurgency operations that emerged during the Vietnam conflict suggested that war might have a second grammar. This one seemed to differ fundamentally from the first primarily because it gave a higher priority to protecting the indigenous populace rather than destroying the enemy. To be sure, the counterinsurgency strain of American military theory – the 'clear, hold, and build' approach – did consider the destruction of insurgent forces a priority, just not the first priority.[9] If the U.S. military sought to define the Vietnam experience as an aberration, as the histories of the post-Vietnam era suggest, it finally succeeded in doing so in August of 1991. Indeed, many contemporary theories suggested a modified role for special operations forces that went beyond training indigenous personnel to defeat insurgents. The term 'global scouts', while not new, gained currency because it captured the sense of using special operations forces to set the conditions for battle.[10]

Desert Storm also revalidated the battle doctrine advanced by airpower theorists, which in brief held that long-range precision-strike technology could be used to target a foe's leadership and will to fight, while avoiding not only costly ground campaigns but also the extensive destruction traditionally associated with strategic bombing.[11] In retrospect, it is clear that too much was made of the fact that the ground phase of the counteroffensive lasted barely 100 hours, as this fact provided grist for a lively debate, a dispute of conflicting battle doctrines.[12] The air phase, by comparison, lasted 39 days, and, for some, appeared to have been almost solely responsible for the successful outcome.[13] For this and other reasons the war was quickly touted as an airpower victory, or more precisely, a victory for long-range precision engagement, and it was offered as proof of the efficacy of fighting wars with smart bombs and cruise missiles rather than large numbers of 'boots on the ground'.[14] Precision engagement, coupled with a nominal ground presence in the form of special operations troops and indigenous forces, was soon hailed as the essence of a 'new' American way of war.[15]

However, the 'boots on the ground' school of thought, represented chiefly by the U.S. Army and U.S. Marine Corps, countered such claims by arguing that any ground force worthy of the name had to be strong enough to compel a decision – to provide a 'forcing function' in military parlance. Otherwise an adversary could resist airpower's attempts at coercion almost indefinitely.[16] To be sure, attacks with smart bombs and cruise missiles could certainly aid in modifying an adversary's behaviour. Nonetheless, it was not clear how often such weapons would suffice. Bad weather, for instance, grounded NATO aircraft frequently enough in the Balkan conflicts to underscore the argument that the concept of friction was still alive and well, even in the so-called virtual wars of the information age.[17]

It was this debate, this duel between battle doctrines, in fact, though hardly new, that shaped how the services' modernization efforts proceeded, and in particular how the revolution in military affairs (RMA) unfolded in the United States. Intra-service debates can prove beneficial since they force each of the services to justify its views concerning roles and missions, and because they help expose the assumptions underpinning competing bodies of theory. One perspective rested on strategic bombing doctrine, the initial ideas of which emerged before the First World War. A second perspective grew out of the manoeuvre school of thought, which could trace its roots to any one of history's decisive battles. The dispute between these two intensified throughout the 1990s. The downsizing of the U.S. military by nearly one third, as recommended by Defense Secretary Les Aspin with the 'Bottom-Up-Review' in 1993, contributed to a marked increase in tensions among the services.[18]

In the Bosnian and Kosovo conflicts land power again appeared to play a limited role. This in turn lent further credence to the claims of airpower enthusiasts, who began to argue, with some justification, that airpower gave America an 'asymmetric' advantage that allowed wars to be fought more cheaply, more quickly, and from a distance. However, the Balkan conflicts were hardly clear-cut in terms of how effective either air or landpower was individually. As one scholar noted: 'Despite precision bombing, enemy decapitation has not become the "new American way of war". Rather, precision weaponry has revolutionized contemporary warfare by multiplying the effectiveness of using air and ground power together.'[19]

To be sure, the Balkan conflicts did pose different challenges to both air and ground commanders, and not only because of the kinds of motives involved, to include a desire to 'ethnically cleanse' certain regions and to achieve, or consolidate, political sovereignty over them.[20] However, they were only the most visible of such operations, which have continued to increase in number since the end of the Cold War.[21] In short, the fact that these conflicts saw 'systematic violence against non-combatants elevated to the status of a strategy', pointed to the shortfalls of a battle-focused doctrine.[22] As in any peacekeeping or humanitarian assistance operation, the key is to establish safe zones wherein some level of political and economic stability can be built.[23] The doctrine for such missions did exist, though training for them received less priority, which is hardly surprising, given the larger mission of deterring the Warsaw Pact. Nonetheless, the 1993 version of FM 100–5 emphasized the importance of battle, even though it acknowledged that the Cold War had ended and the nature of the threat had changed.[24] The principal role of the U.S. Army in the post-Cold War era was deterrence and the ability to project power anywhere; it also made a distinction between three operational environments: peace, conflict, and war. This acknowledgement was not unlike the statement by USMC Commandant, General Charles C. Krulak, about a 'three-block war', meaning that soldiers might have to participate in a decisive military action, conduct peacekeeping operations, and humanitarian relief activities all at the same time, or within a geographically limited area.[25] However, General Krulak addressed in much more detail the leadership implications for such activities. Moreover, in each of these conflicts, rules of engagement (ROE) began to take on unprecedented importance, from the actions of individual soldiers to restrictions on the use of long-range precision munitions. Signals were growing stronger, in other words, that a second grammar of war was needed.

The RMA: perfecting war's grammar

The fundamental differences underpinning ground and air battle doctrines were brought out in two symposia hosted in Washington, D.C., in 1997 and in 1998 by the Centre for Strategic and International Studies.[26] Predictably, the symposia ended without resolving any of the issues. Nonetheless, the point is that it was, indeed, this running battle, taking place as it did in an era of deep cuts that, more than anything else, helped provide the context for the RMA. While all the services were reeling, the U.S. Air Force, in particular, felt the need to justify the high costs of most of its weapon systems, particularly after the recommendations of the *1997 Quadrennial Defense Review*, which proposed that the service be reduced by nearly 27,000 personnel and that its aircraft programs, including the much vaunted F-22, be cut substantially, or halted and consolidated.[27] The way ahead seemed clear: marrying precision strike with information dominance offered a smaller, lower-cost, lower-collateral-damage, yet far-reaching military alternative to labour-intensive ground wars.

If the general downsizing of the services made the threat to 'revolutionize' or restructure the U.S. military appear to be a real one, the move toward greater 'jointness' was supposed to underscore each service's vital contribution to mission accomplishment. While the practical foundation had been laid by the 1986 Goldwater–Nichols Act, substantial conceptual gaps remained.[28] *Joint Vision 2010* (1996) and *Joint Vision 2020* (2000) not only called for U.S. military forces to become 'fully joint' in response to force-sizing pressures and the anticipated demands of the future security environment, they attempted to provide a conceptual basis for doing so.[29]

However, they offered no genuine overarching concept around which the services might 'integrate' themselves. *JV 2010* and *JV 2020* described four concepts which were to set a course for the further modernization of U.S. forces; those concepts were: dominant manoeuvre, precision engagement, focused logistics, and full dimensional protection, all of which were to lead to the objective of full spectrum dominance. The concepts themselves were little more than tautologies, however. Dominant manoeuvre is simply the capability to conduct manoeuvre that dominates; precision engagement is just the capability to engage with pinpoint accuracy. Similarly, full spectrum dominance, in essence, meant that U.S. forces were simply expected to excel across the entire spectrum of conflict. This aim was not only wholly impractical, it failed to bring together two of the four supporting concepts, dominant manoeuvre and precision engagement, which had developed from competing traditions. In short, neither *JV 2010* nor *JV 2020* did more than mollify service interests.[30]

Much evidence suggests that the American idea of institutionalizing the RMA – rather than just letting it run its course – was drawn from the Soviet concept of a military technical revolution, or MTR. For leading defence intellectuals, such as Andrew Marshall and Andrew Krepinevich, the concept of an MTR did not go far enough to capture the extent of change that was possible, or necessary, with information technologies.[31] It was not too difficult, nor entirely incorrect, to cast the history of warfare as a series of revolutions, each in its turn grasped only by those with vision.[32] The idea of an impending revolution might be the only thing that could get the defence department to overcome its infamous inertia and effect meaningful change.

The prevailing view underpinning the RMA movement was that information technologies were indeed transformative enough to lead to a revolutionary change in the way wars were fought. The value of information or knowledge in warfare was self-evident to all the services, as their efforts at modernization indicate. Examples of these include the *Force XXI* initiative for the U.S. Army, which gathered momentum between 1992 and 1995; it was to experiment with various information technologies to project a digitally equipped, integrated ground force that would be fielded between 2000 and 2010. The guiding document for this effort was *TRADOC Pamphlet 525–5*, which described a knowledge enriched, situational aware fighting force employing a variety of robotics and sensor-shooter linkages.[33] In contrast the Army After Next Project, which was established in 1995, was supposed to look at what the U.S. Army's needs might be in the period around 2020, and to link its findings back to *Force XXI* helping it to remain on course. These efforts, just like those fuelling the development of the U.S. Air Force's current array of Aerospace Expeditionary Forces (AEFs), were driven by the desire to optimize the use of information technology and to avoid becoming strategically irrelevant.

However, the new concepts that emerged were often viewed as old wine in new skins, that is, as one service or another attempting to put forth its own preferred modes of thinking and operating. One issue, for instance, was the extent to which any information technology could truly overcome the fog of war.[34] The fog of war – or more properly the friction of war if one is referring to Clausewitz's concept – was seen as an unavoidable aspect of war's nature. As such it provided a fundamental argument for 'big battalions', or having sizeable forces in reserve in the event things went wrong, as they often do. If information could lead to genuine situational awareness, large reserves would no longer be necessary, at least in theory. A secondary issue was whether a balanced joint force was truly what the United States needed, or desired. Balance, despite ample defence rhetoric to the contrary, lies in the eye of the beholder.

By 1999, the idea of an information-linked fighting force, capable of sharing knowledge and thus of responding with greater efficiency and flexibility, had become accepted wisdom. Known as Network-Centric Warfare (NCW), the idea was vigorously advanced by the late Admiral Arthur K. Cebrowski of the Office of Force Transformation.[35] The central assumption was that a network that could receive the right information, whenever it needed to, could increase efficiencies, reduce response times, and enable a smaller number of platforms to accomplish a greater number of missions. Yet a network of obsolete or inappropriate systems can be efficient without being effective. The underlying issue remains what systems one should buy and how many. In short, NCW amounts to little more than ensuring one's systems can communicate with each other, which is not a small item, particularly when one considers the potential participation of allies and coalition partners in any conflict.[36]

One significant concept enabled by NCW was Rapid Decisive Operations (RDO).[37] Its developers clearly intended to use it to put a very high mark on the wall in order, at least in part, to stimulate research and development. According to its White Paper (dated 9 August, 2001), RDO was intended to function as an integrating concept oriented on high-end, small-scale contingencies in the 2015 time frame, and purportedly built upon the five concepts outlined in *Joint Vision 2020.* The idea was to 'integrate knowledge, command and control, and effects-based operations to achieve the desired political/military effect'. As the White Paper went on to say 'military acts' would occur 'in concert with and leverage the other instruments of national power to understand and reduce the regional adversary's critical capabilities and coherence'.[38]

Nonetheless, while RDO called for integrating military action with the other elements of national power, little research was actually directed at attempting to understand how such different kinds of power might be integrated, especially considering that they operate at different rates, and put in motion different chains of effects. Instead, the idea of using 'all the elements of national power' became something of a mantra, implying that in this way every war could be won, or every major obstacle overcome. In any case, RDO was quickly abandoned, with the joint community explaining that it had achieved its primary purpose in establishing other concepts, such as Effects-Based Operations, or EBO.

EBO was supposed to shift focus from achieving 'objectives' to achieving 'effects', which for many respected authorities was a distinction without a difference.[39] A number of EBO enthusiasts even claimed that it was possible to predict, and thus achieve, first, second, third, and fourth-order effects, while minimizing undesired effects.[40] However, EBO came under fire almost immediately from influential officers in the U.S. Army and U.S. Marine Corps for being a thinly veiled attempt by airpower enthusiasts to impose a straightforward, point-and-shoot targeting mentality on a messy, non-linear three-block war environment. Nonetheless, EBO made rapid headway in defence circles. Even though it had not been approved as official doctrine, it began to appear in doctrinal publications and was being taught at staff colleges. Finally, in August 2008, USMC General James Mattis, commander of U.S. Joint Forces Command, issued a memo rejecting EBO precisely (but not only) because the predictability it promised ran counter to his own personal experiences in war, and the history of warfare itself.[41] Although EBO's demise came late, it was telling evidence that rather than perfecting the same grammar of war, elements of the U.S. military had begun to look toward the merits of other types of grammar; but that is a story for another day.

In conclusion, the late historian Russell Weigley once wrote that the period of the 1950s and 1960s, when the pressures of the Cold War forced new ways of looking at armed conflict, amounted to a 'golden age' in American strategic thinking.[42] It is not to be expected that thinkers like Marshall, Owens, and Cebrowski can be compared to Schelling, Osgood and Brodie, or even Kahn. Such comparisons would do little justice to either group.

Nonetheless, it is clear that American military theory once again came into its own, putting 'marks on the wall', as it were, just as it did during its so-called golden age. Part of doing so entailed rejecting some of the Cold War assumptions that underpinned strategic thinking in the nuclear age, and returning to a preoccupation with battle. Part of it also had to do merely with timing: matching a rapidly maturing branch of technology with intra-service competition in a downsizing era. In part, the abundance of theories also depended on defining very loosely, as this author has done, what qualifies as theory.

The revolution in information technology is hardly over. However, it is quickly being subsumed by overlapping and accelerating revolutions in biotechnology, nanotechnology, and related fields. Unfortunately, the thinkers who will influence this period of strategic thinking have yet to take centre stage.

Notes

1 For purposes of this essay, the term *theory* means not only nascent concepts and speculation, which lack the analytical rigour that distinguishes proper theory, but also doctrine, which is theory that bears the stamp of official sanction.

2 Dima Adamsky, 'American Strategic Culture and the U.S. Revolution in Military Affairs', *Defence and Security Studies,* 1:36, discusses several sources which make this argument: Chris Donnelly, *Red Banner: The Soviet Military System in Peace and War,* London: Jane's Information Group, 1981, p. 201; Andrew May, *The Sources of the U.S. Military Advantage,* McLean, Virginia: Science Applications International Corporation, 2002; Richard Lock-Pullan, *U.S. Intervention Policy and Army Innovation,* New York: Routledge, 2006; Donn Starry, 'A Perspective on American Military Thought' in *Military Review* 69, pp. 2–11, 1989.

3 The United States and its coalition partners marshalled some 680,000 troops, 3,000 tanks, 2,000 aircraft, and 100 warships to oppose Saddam Hussein's 336,000 troops and 9,000 tanks. Brigadier General John S. Brown, 'The Maturation of Operational Art: Operations Desert Shield and Desert Storm', in Michael D. Krause and R. Cody Phillips, (eds), *Historical Perspectives of the Operational Art,* Washington D.C.: Center of Military History, 2005, pp. 439–82.

4 Department of the Army, *FM 100–5 Operations,* August 20, 1982, Washington D.C.: GPO, 1982, pp. 1–2. See also Swain, 'Filling the Void', p. 159, and Allan English, 'The Operational Art' in Allan English *et al.,* (eds), *The Operational Art: Canadian Perspectives, Context and Concepts,* Kingston, Ontario: Canadian Defence Academy, 2005, p. 16.

5 English, 'Operational Art', p. 16.

6 Shimon Naveh, *In Pursuit of Military Excellence: The Evolution of Operational Theory,* London: Frank Cass, 1997, pp. 292–5.

7 See: Robert E. Osgood, *Limited War: The Challenge to American Strategy,* Chicago: University of Chicago, 1957; and *Limited War Revisited,* Boulder: Westview, 1979; Thomas C. Schelling, *Arms and Influence,* New Haven: Yale University, 1966; Bernard Brodie, *Strategy in the Missile Age,* Princeton: Princeton University, 1965; and *War and Politics,* New York: Macmillan, 1973; Herman Kahn, *On Thermonuclear War,* Princeton: Princeton University, 1960; and *On Escalation,* New York: Praeger, 1965.

8 Harry G. Summers, Jr., *On Strategy: A Critical Analysis of the Vietnam War,* Navato, Ca.: Presidio, 1995.

9 The following capture some of that body of thought: Franklin Mark Osanka (ed.), *Modern Guerrilla Warfare: Fighting Communist Guerrilla Movements, 1941–1961,* Glencoe, Il: Free Press, 1962; Walter Laqueur, *Guerilla Warfare: A Historical and Critical Study,* Boulder: Westview, 1984; Ian F. W. Beckett and John Pimlott (eds), *Armed Forces and Modern Counter-insurgency,* New York: St. Martin's, 1985.

10 For an example, see *Knowledge & Speed: The Annual Report of the Army After Next Project,* Ft. Monroe, VA: U.S. Army Training and Doctrine Command, 1997.

11 U.S. Air Force Colonel John A. Warden III became the most well known of airpower theorists. In brief, Warden's theory was to attack the critical points in each of five 'rings' simultaneously: leadership, organic or system essentials, infrastructure, population, and fielded military forces. Ideally this would bring about the strategic paralysis or psychological collapse of one's opponent. John A. Warden III, *Air Campaign: Planning for Combat,* Washington D.C.: National Defense University, 1989; see also John Andreas Olsen, *John Warden and the Renaissance of American Air Power,* Washington D.C.: Potomac, 2007.

12 Lawrence Freedman and Efraim Karsh, *The Gulf Conflict, 1990–1991: Diplomacy and War in the New World Order,* Princeton: Princeton University, 1991, pp. 316–17; Anthony H. Cordesman, 'Persian Gulf War (1991)' in *United States History,* 588.

13 Richard P. Hallion, *Storm Over Iraq: Air Power and the Gulf War,* Washington, D.C.: Smithsonian Institution Press, 1992; Rebecca Grant, *Airpower and the Total Force: The Gift of Time,* Arlington, VA: IRIS Independent Research, 1998.

14 D.A. Deptula, 'Firing for Effect: Change in the Nature of Warfare', in *Defense and Airpower Series*, 1995.

15 General Ronald Fogelman, USAF, 'Airpower and the American Way of War', speech presented at the Air Force Association's Air Warfare Symposium, Orlando, Florida, February 15, 1996; see Grant T. Hammond, 'The U.S. Air Force and the American Way of War' in Anthony D. McIvor (ed.), *Rethinking the Principles of War*, Annapolis: Naval Institute Press, 2005, pp. 109–25; see also the presentation by Major General Charles D. Link, 'Responding to Aggression: Boots on the Ground vs. Technology', October 29, 1997, presented at the CSIS *Clash of Visions symposium* (see below).

16 David E. Johnson, *Learning Large Lessons: The Evolving Roles of Ground Power and Air Power in the Post-Cold War World*, Santa Monica, CA: RAND, 2007, discusses briefly how the concerns evolved.

17 Michael Ignatieff, *Virtual War: Kosovo and Beyond*, New York: Metropolitan, 2000.

18 Total active duty personnel went from 2.1 million to 1.4 million; the U.S. Army went from 18 divisions to ten; the U.S. Air Force went from 24 active fighter wings to 13; the U.S. Navy went from 567 ships to 346; and the U.S. Marines went from 197,000 personnel to 174,000. Cf. Earl H. Tilford, Jr., *The Revolution in Military Affairs: Prospects and Cautions*, Carlisle: Strategic Studies Institute, 1995, p. 3.

19 Robert A. Pape, 'The True Worth of Air Power' in *Foreign Affairs* 83:2, March/April 2004, pp. 116–30.

20 Mary Kaldor, *New & Old Wars: Organized Violence in a Global Era*, Stanford: Stanford University, 1999.

21 This mission set is discussed in Huba Wass de Czege and Antulio J. Echevarria II, 'Toward a Strategy of Positive Ends', in *National Security Studies Quarterly* 8:1, Winter 2002, pp. 1–26.

22 R. Craig Nation, *War in the Balkans, 1991–2002*, Carlisle Barracks: U.S. Army War College, 2003, p. ix.

23 In 1990, as the Cold War ended, the United Nations had five peacekeeping operations underway, excluding those on the Korean peninsula, involving about 10,000 troops; whereas in 2006 it had 18 such operations taking place with nearly 73,000 troops involved. Peacekeeping costs had risen from $800 million in 1990 to $41 billion in 2006. Cf. United Nations Peacekeeping Operations, Background Note, dated February 28, 2006. Online. Available at: www.un.org/depts/dpko/dpko/bnote.htm.

24 Department of the Army, *FM 100–5 Operations*, June 14, 1993, Washington, D.C.: GPO, 1993, pp. 1–3; David A. Fastabend, 'Checking the Doctrinal Map: Can We Get There from Here with FM 100–5?' in *Parameters*, Summer 1995.

25 USMC Gen. Charles C. Krulak, 'The Strategic Corporal: Leadership in the Three Block War' in *Marines Magazine*, 1999. Online. Available at: www.au.af.mil/au/awc/awcgate/usmc/strategic_corporal.htm.

26 The first symposium, *Clashes of Vision*, took place in October 1997; the second one, *Dueling Doctrines and the New American Way of War*, occurred in June 1998. Cf. Earl H. Tilford, Jr., *Halt Phase Strategy: New Wine in Old Skins. ... With Powerpoint*, Carlisle, PA: U.S. Army War College, 1998, p. 3.

27 See the summary in Tilford, *Halt Phase Strategy*, p. 2.

28 Gordon Nathaniel Lederman, *Reorganizing the Joint Chiefs of Staff: The Goldwater – Nichols Act of 1986*, Westport, CT: Greenwood Press, 1999; James R. Locher, *Victory on the Potomac: The Goldwater – Nichols Act Unifies the Pentagon*, College Station, TX: Texas A&M University Press, 2002.

29 *Joint Vision 2020*, Washington D.C., June 2000, Internet copy, p. 2; *JV 2010* is the forerunner of *JV 2020* and essentially discusses the same concepts.

30 Likewise, the *Concept for Future Joint Operations: Expanding Joint Vision 2010*,

Washington D.C., May 1997, states that 'Joint operations require a common frame of reference and guiding principles to integrate and focus joint capabilities toward common objectives', p. 61; but it does little more than offer several pages of new definitions.

31 Adamsky, *American Strategic Culture*, pp. 26–9, provides an excellent summary, and is especially valuable on the Soviet perspective.

32 Andrew Krepinevich, 'Cavalry to Computer: The Pattern of Military Revolutions' in *National Interest*, 37, Fall 1994; and 'Keeping Pace with the Military Technical Revolution' in *Issues in Science and Technology*, Summer 1994; for a rebuttal see Tilford, *The Revolution in Military Affairs*. For other views see Dan Goure, 'Is There a Military–Technical Revolution in America's Future?' in *Washington Quarterly* 16:4, Autumn 1993, pp. 175–91; James R. Fitzsimonds and Jan M. Van Tol, 'Revolutions in Military Affairs' in *Joint Force Quarterly*, Spring 1994, pp. 21–31; Eliot Cohen, 'Come the Revolution' in *National Review* 47:14, July 1995, pp. 26–31; 'A Revolution in Warfare' in *Foreign Affairs* 75:2, March/April 1996, pp. 37–45.

33 *Pamphlet 525–5, Force XXI Operations*, Fort Monroe, VA: U.S. Army Training and Doctrine Command, 1994.

34 Admiral Bill Owens, *Lifting the Fog of War*, New York: Farrar, Strauss, and Giroux, 2000; see also 'The Emerging U.S. System of Systems' in Martin Libicki and Stuart Johnson (eds.), *Dominant Battlespace Knowledge: The Winning Edge*, Washington, D.C.: National Defense University, 1996.

35 Vice Admiral Arthur K. Cebrowski and John J. Garstka, 'Network-Centric Warfare: Its Origin and Future' in *Proceedings of the U.S. Naval Institute*, January 1998; Arthur K. Cebrowski, 'Network-Centric Warfare: An Emerging Military Response to the Information Age', presented at the 1999 Command and Control Research and Technology Symposium, June 29, 1999; see also David S. Alberts *et al.*, *Network Centric Warfare: Developing and Leveraging Information Superiority*, Washington D.C.: C4ISR Cooperative Research Program, 1999, p. 2, who define NCW as:

> an information superiority-enabled concept of operations that generates increased combat power by networking sensors, decision makers, and shooters to achieve shared awareness, increased speed of command, higher tempo of operations, greater lethality, increased survivability, and a degree of self-synchronization.

36 David C. Gompert *et al.*, *Mind the Gap: Promoting a Transatlantic Revolution in Military Affairs*, Washington, D.C.: National Defense University, 1999.

37 'Rapid Decisive Operations (RDO) White Paper', United States Joint Forces Command (JFCOM), J9 Joint Futures Lab, Coordinating Draft (Version 2.0), dated August 9, 2001. RDO synthesizes a broad range of ideas extracted from works by RAND, the Institute for Defense Analyses, Defense Group, Inc., and others.

38 The White Paper further states:

> The United States and its allies *asymmetrically* assault the adversary from directions and in dimensions against which he has *no counter*, dictating the terms and tempo of the operation. The adversary, suffering from the *loss of coherence* and unable to achieve his objectives, chooses to cease actions that are against U.S. interests or has his capabilities defeated.
>
> (RDO, p. ii; emphasis added.)

39 The official definition of EBO is:

> a process for obtaining a desired strategic outcome or "effect" on the enemy through the synergistic, multiplicative, and cumulative application of the

full range of military and other national capabilities at the tactical, operational, and strategic levels.

(Rapid Decisive Operations White Paper, January 2002, p. 20.)

The USAF Air Combat Command's (ACC) EBO White Paper states that EBO is a mere refinement of objectives-based planning; but why such refinement was necessary was not made clear.

40 David Deptula, *Effects-Based Operations: Changes in the Nature of Warfare*, Arlington VA: Aerospace Education Foundation, 2001.

41 General J. N. Mattis, USMC, Commander U.S. Joint Forces Command, 'Commander's Guidance Regarding Effects-Based Operations', Memorandum, dated August 14, 2008; and 'USJFCOM Commander's Guidance for Effects-based Operations', in *Parameters* 38:3, August 2008, pp. 18–25.

42 Russell F. Weigley, *The American Way Of War: A History Of United States Military Strategy And Policy*, Bloomington, IN: Indiana University Press, 1973.

6 The revolution in military affairs with Chinese characteristics

Jacqueline Newmyer Deal

Introduction: variation in RMA perspectives

A fundamental change in the behavior of military organizations and the conduct of war raises the question of which states or actors are best positioned to benefit. It has been demonstrated that the Soviet Union saw the contemporary revolution in military affairs (RMA) as delivering a major advantage to the United States. This chapter argues that a traditional Chinese strategic outlook emphasizing superior information, intelligence, and the manipulation of perceptions to prepare the battlefield in peacetime shapes the People's Republic of China's (PRC's) approach to the RMA. The advances in computing and communications and the fundamental shift in strategic affairs associated with the RMA therefore provide China, at least in Beijing's eyes, with an opportunity to benefit disproportionately relative to its rivals. In particular, having long downplayed the People's Liberation Army's (PLA's) capabilities and aims, China is now publicizing its capacity to inflict damage while deliberately cultivating uncertainty about its precise intentions as a means of inducing caution among its adversaries. Key dimensions of China's RMA include complementary, serialized electronic and kinetic attacks and the addition of 'information deterrence' to nuclear deterrence. A related finding is that today's environment differs fundamentally from the Cold War context in which the RMA first emerged.[1]

The specter of a confrontation between NATO and Warsaw Pact forces in Europe formed the backdrop for the developments associated with the Soviet identification in the mid- to late-1970s of an American-led 'military-technical revolution', subsequently known as the RMA or simply 'military revolution' in the United States.[2] This fact may constitute the lone point of agreement within a set of lively RMA debates among strategic studies scholars and students of the Cold War. In particular, notwithstanding controversies about the existence, definition, significance, and future of the RMA,[3] its documentary origins have been traced to the Soviet observation that the United States was exploiting developments in computer processing and other technologies to achieve a 'reconnaissance-strike complex' (RSC) capable of targeting Soviet forces based deep in the rear. While it is clear that the Soviet and American defense establishments exhibited significant variation in the degree to which, and the ways in which, they conceptualized and employed the RMA,[4] the record shows that doctrinal developments in the United States and the USSR were rather tightly coupled in the period of the RMA's birth, with the U.S. Air–Land Battle doctrine a clear response to the Soviet echelons approach. By the mid- to late-1970s the United States and the USSR shared a perspective on what would be the dominant engagement if their competition devolved into a hot war. Their approaches, while different, did proceed from a common point of departure.

Today, by contrast, the nature of the dominant engagement, and even the existence of a competition between the United States and China are disputed. Turning to Chinese writings on the RMA promises to shed some light on why this may be the case.

Chinese writings on the RMA: is imitation the sincerest form of flattery?

What is the Chinese conception of the RMA (*xin junshi geming*, new military revolution, or *junshi geming*, military revolution)? A logical first approach to the question would be to investigate the prevailing way of defining the RMA in the PRC. Which, if any, of the various definitions and understandings within the United States and elsewhere around the world do Chinese defense analysts adopt? Answering this question is complicated by the fact that People's Liberation Army (PLA) strategists who write in the journals of the Chinese Academy of Military Science and National Defense University, among other outlets, tend to follow international military thought closely, as they have inherited a strategic outlook that emphasizes 'knowing the enemy' and carefully monitoring trends (as will be discussed in further detail below). For instance, one of the leading Chinese writers on the RMA, Major General (ret.) Wang Baocun, is also a translator of Paul Kennedy's *Rise and Fall of the Great Powers*.[5]

Perusing Chinese military journals and edited volumes from the 1980s and 1990s yields an impressive array of definitions that seem to mimic lines from Russian and American sources – lines that emphasize organizational and doctrinal shifts associated with technological advances allowing for dramatically enhanced reconnaissance and precision at increasing ranges. The PLA watched and learned from a distance as the United States employed new RMA capabilities first in Operation Desert Storm and then in the wars in the Balkans in the 1990s. The Chinese were reading Ogarkov's work, studies commissioned by the Office of Net Assessment, and other Western analysis,[6] and in many cases it is difficult to discern whether they added a particular gloss of their own to the foreign assessments.

But in this period one can also find articles emphasizing the new vulnerabilities and opportunities for information warfare (IW, or *xinxi zhan*, alternately translated as informatized or informationalized [*xinxi hua*] war) created by the reliance of militaries, as well as broader social and economic systems, on computer networks. These discussions seem to reflect unique Chinese contributions, or a unique synthesis of Russian and American writings that ends up looking like neither.

Consider, for instance, this article by then Senior Colonel Wang Baocun, published in the *PLA Daily* newspaper in April 1998, worth quoting at length:

> The opportunity created by the new military revolution is a chance of a lifetime.... Our country has achieved rapid progress in informatization and has the basic 'potential energy' to extend this work to the military. An important feature of the present military revolution is that local informatization begins sooner and develops faster than in the armed forces and is technologically more advanced. After building sufficient 'potential energy', the work will then be extended to the military and will trigger off an enormous military transformation....
>
> The potential of information technology to diffuse extensively provides the technological conditions for the informatization of the army. A direct driving force of the present military revolution is the rapid development of high technologies, especially information technology.... Unlike nuclear and stealth technologies, information technology has greater potential for diffusion and penetration and is not easy to keep secret. For example, information technology has military and civilian applications, and advanced information technology products are available in international markets. In a sense, the information society shaped by information technology is a "global village" without national boundaries, in which the flow of technological information is not only swift but unstoppable. Tools of information technology carry information, and information has no value unless it is disseminated. For this reason, if we can do a good job in absorbing the advanced information technology of other countries, we can speed up the informatization of the army.[7]

While some of the arguments about the Internet and characteristics of the RMA in this piece overlap with points made in a 1996 *Foreign Affairs* article by Eliot Cohen,[8] Wang's method of exposition and the conclusions he draws for China are *sui generis*. Without articulating a threat or raising cause for alarm, he has pointed out that the PLA is positioned to appropriate the fruits of research and development in other countries and thereby to 'leap' into a dominant military position.[9] Indeed, the goal of reaping competitive benefits from open trade and technology flows creates an imperative to reassure the United States and other militarily advanced states. Hence the omission of any specific platforms, either American or desired Chinese, in favor of statements like this:

> During the Eighth Five-Year Plan period, China's telephone switching capacity increased by 58.99 million lines, bringing the total interoffice switching capacity to 71 million lines and the total switching capacity of urban and rural telephones to 85.10 million lines. China thus became a country with one of the largest telephone networks in the world.

And this: 'If we take the matter lightly and let the opportunity slip past, we will once again be discarded by history when developed countries have completed their work.'[10] Wang thus uses the specter of national humiliation to make the case for leaping into first place among all military powers.

To prevent the abstract language and invocation of China's historical deficits from distracting us from the PLA's practical efforts to leapfrog, it helps to keep in mind what China was pursuing in the way of both electronic and kinetic capabilities at the time. For instance, in the years running up to Wang's publication, the periodical *Naval and Merchant Ships* (*Jianchuan Zhishi*), published by the Chinese Society of Naval Architecture and Engineering (CSNAME), a major Chinese shipbuilding concern, had run a series of articles on technical aspects of naval operations ranging from the uses of infrared sensors on naval attack planes and helicopters and methods for jamming anti-ship missiles to anti-submarine warfare (ASW) acoustics, various types of sonar arrays, and stealth casings for torpedoes to defeat sonar.[11] Experts in the United States also trace China's pursuit of the DF-21D anti-ship ballistic missile (ASBM) to the Taiwan Strait Crisis of 1995–1996, during which President Clinton deployed a U.S. carrier group to dissuade Beijing from conducting additional missile firings across the Taiwan Strait. Where Air–Land Battle was focused on a contingency in the heart of Europe, Chinese RMA doctrine seems to incorporate a heavy naval component. The RMA enables many of the capabilities we now associate with China's 'anti-access, area denial' (or, to use the Chinese term, 'counter-intervention') effort to keep U.S. forces well outside not just the Taiwan Strait but all the so-called near seas – i.e., the Yellow Sea, the East China Sea, and the South China Sea. More broadly, the use of electronic warfare and guided missile strikes against critical command-and-control platforms – as well as softer fixed targets like runways or ports – figures prominently in emerging Chinese doctrine. The goal seems to be to isolate enemies operating over significant distances in the air, on land, or at sea.

The Chinese RMA has applications not just at the tactical or operational level but also at the strategic level, as outlined in a 2001 article by Zhou Fangyin in the *Journal of the Chinese Institute for Contemporary International Relations* (CICIR), an arm of the PRC's intelligence/counterintelligence apparatus. This article is worth citing at length because it is representative of the open-source Chinese RMA literature. The excerpt below begins with Zhou's third point, following a discussion about how the increased transparency of political actions has reduced the scope of warfare (point one) and its destructiveness (point two), which may again be taken as a form of reassurance:

> 3. Transformation from nuclear deterrence to information deterrence
>
> Traditional deterrence theory is mainly nuclear deterrence theory. Nuclear weapons are capable of enormous destruction far exceeding that of conventional weapons. Thus, nuclear weapons can produce a huge social and psychological reaction, and they have a unique deterrent effect. The core of nuclear deterrence theory is to "use the non-use" of nuclear power as a means to force an enemy to abandon the launching of a nuclear offensive or other warlike action, and thus achieve a nation's political, security, and military objectives. Therefore, some people believe that a "nuclear weapons umbrella" in a sense serves to protect security in the nuclear age.
>
> The concept of an "information umbrella" is an extension of the concept of a nuclear weapons umbrella.... A country which has information superiority can use it to influence the behavior of other countries.... Although an information umbrella does not present a fearsome threat, it has much greater potential for use than a nuclear weapons umbrella.
>
> The basic meaning of an "information protection umbrella" is to unfold an electronic "umbrella" capable of maintaining information superiority over the top of a country engaged in battle, to enable us to see the opponent without the opponent seeing us, to enable us to protect ourselves effectively while the opponent cannot protect himself, to enable our side to act more quickly while the opponent does not have time to react, and thus to form up deterrence against the opponent or to subdue him.
>
> Any form of military attack can under certain conditions become a form of deterrence.... Holding information superiority can accomplish two objectives. The first is to use information deterrence to force an adversary to lay down his weapons. In this way, a military objective can be achieved while avoiding losses and casualties on both sides. Second, ... so long as one's own side has information superiority, it can execute precise attacks on the opponent. In this situation, the side which holds information superiority can frequently make a huge strike on the opponent at an extremely small price, and ultimately be victorious in the war.

Having established the concept of a 'information umbrella' with deterrent power analogous to that of a 'nuclear umbrella', Zhou goes on to describe the operational uses of information superiority. These uses extend from peacetime disruption of an adversary's civilian information technology infrastructure to wartime disruption of the enemy's command-and-control and targeting capabilities. In a section of the article that specifically mentions research by the U.S. Defense Advanced Research Projects Agency (DARPA), Zhou explains:

> 6. Information security prompts a great deal of attention
>
> The main content of information war in its present stage is electronic warfare and attacks on computer networks. It also involves other related combat operations, such as the use of precision guided weapons and ordinary weapons to attack military information systems .... In information warfare, contests and confrontations revolving around information are certainly not limited to the military sphere. They also involve social, political, economic, and technical aspects. Moreover, the measures used in information warfare are varied. Some measures are violent, and others are nonviolent.
>
> ... Future combat operations in networks will take two main forms, one being computer viruses, and the other network 'hackers'.
>
> In the military sphere, two types of weapons and equipment are most susceptible to infection and damage by computer viruses. One is various kinds of military information systems, such as C4I systems, computer networks, radar systems, and various kinds of sensors. The other is various kinds of high-tech weapons systems which are controlled by computers, such as the automatic drive, fire control, and guidance systems of modern aircraft, ships, tanks, missiles, and other equipment
>
> ... Information networks with computers as their core are the nerve centers of modern armed forces. How well or badly they perform, and how good or bad their condition, has a direct influence on an armed force's combat power and on the operations of a country's entire military machine. Because of this, information networks will play an extremely important role in future combat operations. In the information age, a country's activities in the political, military, economic, social, and scientific and technological spheres will rely more and more on information and computer networks. Computer networks of various kinds are becoming a country's strategic lifeline. Information security is thus becoming more and more of a critically important element of national security.[12]

A number of exotic terms confront American readers of this piece by Zhou, Wang's earlier article, and other work by Chinese strategists – despite the authors' use of the United States as a model or benchmark. Phrases such as 'potential energy' (what China will store up as it develops high-tech capabilities), 'warfare engineering' (the use of simulations and other peacetime activities to determine conflict outcomes), and 'paralysis combat' (the use of threats to or attacks on information infrastructure to paralyze the enemy) cited above, as well as others like 'assassin's mace' (a secret weapon that can enable the inferior to defeat the superior) and 'invisible forces' (communications and other high-tech capabilities that are not as easy to count up as guns and tanks), highlight Chinese attention to matters that have been, at best, at the periphery of American military thinking in an era of nationbuilding, counterterror, and counterinsurgency campaigns.[13] Specifically, it seems that Chinese defense analysts have conceptualized the RMA as a set of technological advances that create new opportunities to target an enemy's resolve through the threat or infliction of a series of focused, limited, but highly damaging strikes. The adversary will be left wondering what other capabilities have been held in reserve. In a world in which nuclear weapons raise the specter of total destruction, cyber attacks and precision strikes may be employed to generate acute pain or losses and to deter further escalation.

In this vein it is noteworthy that when Chinese military researchers invited a small group of American analysts to Beijing in March 1998 for one of the first post-Tiananmen U.S.–China defense gatherings, albeit at the sub-official level, it was the RMA that was on the agenda.[14] During the meeting the Chinese speakers focused on the impact of the high-tech aspects of the RMA on command-and-control, specifically asking the Americans about the use of simulations, 'the organization of U.S. divisions and the flexibility of the division commander to locate himself at different command posts, and how easily he could communicate with subordinate units and headquarters staff officers', as well as the U.S. Defense Department's 'use of IT [information technology] to protect its resources and the use of networks to segregate information traffic'.[15] This suggests a clear focus on the ways in which combat simulations are conducted and influence U.S. decision making; the strengths and weaknesses associated with U.S. command-and-control arrangements in a networked environment; and the protection or vulnerability of data in Pentagon computers. By 2004, according to the Defense White Paper issued by the State Council Information Office that year, the PLA had embraced an official doctrine of an 'RMA with Chinese Characteristics' that was described as having 'informationalization at the core'.[16]

As hinted at above, we may be deceived if we rely only on what is written, particularly in English language publications that Chinese authorities can expect foreigners to read. Even if the extant Chinese writings are not designed to mislead, one must account for the possibility that not all RMA exponents within the PRC will be equipped to determine or to foresee its evolution in China. Research by Michael Pillsbury, for instance, indicates the existence of an RMA constituency within the PLA advocating certain technologies and directions for force transformation in opposition to advocates of older, more traditional 'People's War' and 'Local War' doctrines.[17] (The latter schools emphasize the need to prepare China to absorb and then gradually attrite an invading force, and to defeat another power in a limited, local conflict, respectively.) Stepping back from what has been written, we can try to gage the RMA's impact by considering how it corresponds to or interacts with deeper traditions in China's approach to matters of war and peace.

Chinese strategic culture

From the close monitoring of foreign perspectives to the emphasis on information warfare, many aspects of the Chinese writings on the *xin junshi geming* come into focus when considered in light of the strategic tradition inherited by the PRC. Further, this tradition illuminates some observable contemporary Chinese strategic behavior that may be associated with the RMA even though it is not discussed in the writings – including the acquisition and selective revelation of new Chinese capabilities. What is the connection between the current regime in Beijing's approach to the RMA and China's strategic tradition? The answer lies in the endurance of certain fundamental philosophical and political views that are reflected in the tradition and continue to shape the regime's behavior concerning war and peace. It is no accident that Deng Xiaoping encouraged senior PLA strategists to study the ancient Chinese military classics as he launched them on the course of modernization, or that he compared the contemporary security environment to the world of the Warring States Period, when the classics were written.[18]

The Warring States Period (*c.*450–221 BC), from which emerged China's most famous book on strategy, Sun Zi's *Art of War*, was a founding moment for the Chinese autocratic regime. By the end of the period, the Qin dynasty had finally managed to centralize control over all the lands that then made up the Chinese ecumene, prevailing by outmaneuvering and defeating the six other states with which it had been vying for ascendancy for more than a century. Sun Zi's masterpiece offers stratagems and counsel developed for that struggle.

A key feature of the Warring States context was the performance-based nature of political legitimacy. Rulers were judged on their ability to provide at least a subsistence level of goods and to preside over a stable realm, and verdicts were always rendered retrospectively: a dynast was considered to have lost the right to rule if and when he had failed to survive a challenge.[19] The ruling house of a warring state endured so long as the state's peasants and land owners had confidence in its stewardship, confidence derived from the enjoyment of material comfort rather than any organic allegiance. But in the wake of a natural disaster or when confronted with a dynasty that had succumbed to corruption and was no longer capable of rallying forces to its defense, invaders, rebels, or both were likely to encroach. Underlying the performance-based criteria for legitimacy, the various schools of traditional Chinese philosophy converged in emphasizing the pursuit of harmony with one's environment. Harmony could be achieved through proper social relations and the observance of rites (Confucianism) but also through aligning oneself with nature (Taoism). Both Confucianism and Taoism demand situational awareness, then, as both schools see external signals as the guide for behavior. From this perspective it is not surprising that signs of tumult in the realm were especially troubling, evidence of political malpractice.

This made for internally preoccupied regimes, with strong Warring States rulers deploying informant networks to report on potentially seditious activities. And it encouraged volatility: after gathering intelligence, in the face of a challenge rulers were known to strike out or crack down dramatically, lest the appearance of weakness generate its own momentum and encourage other threats. It is against this backdrop that we must understand Sun Zi's insistence on the need both for attention to trends and, where necessary, bold action, to ensure success at a moment of maximum danger.

It is also critical to recognize certain structural factors about the Warring States Period that were conducive to using peacetime to prepare for war, including the shallowness of alliance relationships and the relative porosity or interpenetration of the various rival states. On the first point, security pacts among regimes with domestic stability concerns proved fragile, as outlying landholders could sometimes be bought off by an invading power, or as a ruling house was convinced that its chances of survival would be enhanced by a change in diplomatic alignment. On the second point, the various warring states existed in close proximity and were mutually intelligible culturally and linguistically. Borders were open, with emissaries often traveling from one capital to another. The states were thus eminently knowable to one another. A ruler who was already dispatching spies within his realm could have confidence in his ability to gather information from his agents abroad. At the same time he had to worry about foreign spies, double agents, and false defectors in his midst, as Sun Zi exhorts. In a world of fluid allegiances, with enemies plotting both within and outside the realm, rulers could not count on a sharp line separating war from peace but rather had to remain vigilant about potential collaboration between foreign and domestic foes.

The remedy prescribed by Sun Zi and the other Chinese classics is to work to construct a secure environment by eliminating enemies and potential rivals starting in the immediate vicinity and building out from there. All conceivable means are included in the arsenal for accomplishing this – from ruses and sabotage to direct attacks – and the fact that enemy rulers and military leaders could be known personally opens up possibilities for exploitation of their particular weaknesses, physical or psychological. By offering inducements, using blackmail, and at times applying deadly force, the texts counsel, a network of friendly or dependent powers can be created as hostile coalitions are divided and weakened. The Qin state that eventually prevailed in the Warring States Period was originally a peripheral one, benefiting from the infighting that occurred among the central powers as it built up its capabilities. In embarking on its conquest, the Qin employed a mix of behind-the-scenes diplomatic maneuvers, covert actions, and well-timed brutal direct attacks.

In sum, the Chinese strategic tradition may be said to present a dynamic, intelligence-based approach to competitions with other powers. Adversaries can be expected not only to mount open challenges but also to plot and encourage subversive activities in one's homeland, so they must be continuously watched and assessed. Further, because a failed military enterprise would endanger the regime's domestic legitimacy, moments for action must be carefully calculated. Force should be deployed decisively, when the grounds have been prepared so that success is virtually guaranteed.

When Deng uttered ancient aphorisms it was easy for Westerners to ignore the unfamiliar references, but consideration of the classical Chinese strategic corpus that he embraced raises questions that compel our attention, especially in light of his role as the sponsor of the PLA's transition from a People's War force to a modernized, RMA-savvy military. For instance, how much of the traditional Chinese approach to politics and legitimacy has endured beneath the Marxist trappings of the PRC? To what degree did Deng perceive the waning of the Soviet Union as ushering in an era of flux and jockeying for power that resembled China's classical founding period? Was he taken with a comparison between the interpenetrated warring states and the porosity of modern states in an era of low mobility costs, peace, and high levels of global commerce? How much should we then read into his famous invocation of the classical Chinese line about 'biding time and hiding capabilities' in 1991?[20] A review of Deng's multi-volume *Selected Works* reveals only one other occasion when he spoke similarly, in a report delivered at a meeting of senior cadres of the Taihang sub-bureau of the Communist Party Central Committee in 1943:

> The task of the underground Party organizations in enemy-occupied areas is to gather strength secretly by every means available and to bide their time. They should try to organize well-selected cadres to work underground as extensively as possible. Party members should try to infiltrate all enemy and puppet organizations, as well as local feudal organizations, to carry out their own activities.[21]

Putting these references from 1943 and 1991 together, one could form an impression of the classical line as a response to difficult circumstances for the Chinese Communist Party. In shepherding China through the period of the fall of the Soviet Union, Deng's approach, as it had been in the early 1940s when the Party faced Japanese invaders and Kuomintang (Nationalist) rivals, was to advocate the maintenance of a low profile in the face of danger, concomitant with the pursuit of the 'strength' necessary to overcome it.

Analysis of China's Warring States strategic tradition, then, suggests that the PRC's approach to the RMA has been guided by a worldview designed to counter an adversary who poses an external military challenge while also threatening internal stability. Given the advanced capabilities with which this foe is endowed in Chinese writings and military exercises,[22] the adversary in question can only be the United States. According to the tradition, the prescription would be to conceal or create uncertainty about China's posture while gathering intelligence and executing military and diplomatic measures to build up forces – preparing the battlefield. The goal, it follows, would be to acquire the capacity to present the United States with a disposition of forces, or, if necessary, a show of force, so menacing as to virtually guarantee the disappearance of a challenge. With this framework in mind, having surveyed the Chinese writings on the RMA, it makes sense to turn to the record of Chinese activities in the RMA era.

The early Chinese response

China's approach to the RMA can be divided into two periods, with a detection and investigation phase (from the late 1980s to the mid- to late-1990s) paving the way for the current implementation phase. Early in the detection phase, consistent with the classic strategic emphasis on intelligence and monitoring trends, the Chinese sought to draw lessons from the end of the Cold War. The conclusion of PLA thinkers like Major General Xu Hezhen was that the USSR owed its defeat to the Strategic Defense Initiative (SDI) and American Information Warfare (IW), with the former draining the economy and the latter sapping the Soviet will.[23] Both the cost-imposing SDI and the 'virus' of democracy are linked to the RMA, as the Chinese understand it to comprehend the acquisition of particular high-tech capabilities like missile defense systems and the use of communications infrastructure to spread ideas damaging to an enemy regime. Xu and other senior PLA officers have written of U.S. efforts to use IW against China.[24]

In terms of practical military effects, the Chinese observed the RMA in action in the first Gulf War and then in Kosovo, where the United States accidentally bombed China's Belgrade embassy, and their descriptions of these campaigns as examples of 'non-contact' and 'informationalized war' should give pause to those inclined to interpret such terms as non-kinetic. Consider, for instance, this passage in an article called 'Military Theoretical Innovation Needed for Preparing for Information War, High-Tech War', co-authored by a professor in the campaign department at China's National Defense University:

> If we say war in the industrial age is 'iron and steel' confrontation complete with imposing arrays of troops, then war in the information age will emphasize the asymmetrical contest of information that is silent and invisible. This trend is hastening the birth of a brand-new form of war. One new form of war is non-contact warfare, which had its debut in the Gulf War and distinguished itself in the Kosovo war. Today it continues to make big strides in the direction of precision, invisibility, and knowledge. To deal with 'non-contact' war, the most important thing is to develop innovative military theories, disengage ourselves from the traditional contact war model, and break new ground in joint operations, in integrated air and outer space warfare, and in information network warfare.[25]

The connection between 'non-contact' and joint, integrated military operations emerges clearly.

Other PLA analyses of the Gulf War and Kosovo similarly expose the links between abstract concepts and the concrete military capabilities that their authors believe China should acquire. Note the interpretation of Kosovo as an informatized war, for instance, in this paraphrased passage from Major General Dai Qingmen, head of an unspecified (suspected IW) department in the headquarters of the PLA General Staff:

> In terms of the concept of the success or failure of informatized war, the goal is to control the enemy and preserve oneself. The objective of controlling the enemy and preserving oneself was exemplified during the war in Kosovo. Here, in 1999, the U.S. military conducted large-scale air raids on Yugoslavia and forced them to surrender under duress without penetrating deep into Yugoslav territory. The success or failure of informatized war is not determined by the ratio of casualties on either side or whether one side has captured the other side's territory, but rather in forcing the enemy to submit to one's will.[26]

In addition to this concept of 'controlling' the enemy, Dai demystifies 'information war' and 'assassin's mace' or 'trump card' (*shashoujian*) weapons with reference to U.S. conduct in the first Gulf War and Kosovo in a 2000 article entitled 'Innovating and Developing Views on Information Operations':

> *Synthesization of arms and equipment for fighting an information war in single-dimensional space is a natural demand of seizing information superiority.* In a future war, a belligerent with information superiority is bound to give scope to its own superiority and try to gain the initiative in operations by making full use of various information fighting platforms in three-dimensional space, including a ground-based platform, a sea-based platform, an air-based platform, and a space-based platform, and

by developing a C4 ISR system, which integrates human functions with mechanical functions and covers all-dimensional space so that every single-dimensional space, such as ground, sea, air, space, and electronics, will become a battlefield where information will be fiercely contended for; every information struggle in single-dimensional space will bear on eventual control of information; and contention for information control in every single-dimensional space will affect a war in terms of process and outcome. Only by launching an attack on an enemy's C4 ISR system in all-dimensional space at the same time or one after another will it be possible for an inferior belligerent to effectively raise overall effectiveness of information operations and seize local information control. For this reason, we should own arms and equipment for fighting an all-round information war so as to contend with an enemy for information superiority in all-dimensional space.

... *Serialization of 'trump cards' and conventional arms and equipment for information operations is a key basis for contending for information superiority.* Information operations are a series of operations with an information environment as the basic battlefield conditions, with military information and an information system as the direct operational targets, and with electronic warfare and a computer network war as the principal forms. Though some high and new technological 'trump cards' among arms and equipment for information operations are likely to play a significant role in information operations, since an enemy may possess a highly integrated information system capable of resisting a powerful information attack, an army armed with a single "trump card" among some powerful arms and equipment for information operations are still likely to find it hard to secure information superiority in most cases.... Only by simultaneously developing and serializing high and new technological arms and equipment for information operations as well as conventional arms and equipment for information operations and making different types of means supplement one another will it be possible to create favorable conditions for gaining more initiative in a war. For this reason, while giving priority to and quickening development of advanced arms and equipment for information operations, including 'trump cards', we should also make a success of research and renovation of conventional information operational means, including conventional arms and equipment, and should serialize information operational means with an eye on effectively contending for information superiority in future.

Finally, not all of the analysis of the United States was so positive. Note the language about an 'inferior belligerent' and 'an army with less advanced arms and equipment' above, and then consider this assessment of the changes in warfare wrought by the RMA, published in the quarterly journal of the PLA Academy of Military Science and the China Military Science Association:

> Since ancient times, there have never been combat operations in which stratagems were not employed. Warfare in different eras has different characteristics, and the role which stratagems have played in wars throughout history has not been the same either. In informationized war, the high degree of complexity in the confrontation in information space provides a broader stage on which to employ stratagems. It could be said that in comparison with other combat actions, actions aimed at seizing information supremacy lay more stress on the use of stratagems. When faced with a powerful enemy, flexible strategies and tactics and innovative methods of combat should be used to compensate for inadequacies in combat power. If our Army hopes to seize information supremacy in the informationized warfare of the future, it must bring its own superiorities into full play, and employ stratagems creatively. As for the commitment of forces, this can be flexible and ever-changing, with advantages concentrated to overcome the opponent's strength. As for the method of combat, speed can be used to defeat strength. Use ever-changing tactics and ever-changing combat dispositions to disrupt the rhythm of a powerful enemy's combat operations, make him unable to completely control the course of his operations, and make it the case that the intelligence which the enemy collects in advance is not an effective guide to his combat actions, or even misleads him.[27]

The Chinese were thus also tracking signs of U.S. weakness in the 1990s. Embedded in theoretical treatises on the future of warfare under informationalized conditions, PLA officers are arguing that the RMA affords China the chance to 'defeat the powerful enemy', as the article later refers to the United States and the United Kingdom.

Further reading of Chinese reactions to the RMA in practice in the 1990s confirms that these interpretations are representative. Published PLA analysis covering the information-gathering phase highlights how the United States was able to achieve victories through well-timed – i.e., surprise – and well-coordinated or 'informatized' strikes. Enemies were not able to strike back at the United States, so the American military mastered 'non-contact' warfare; however, vulnerabilities in the U.S. military and strategic posture were diagnosed. Accordingly, China was said to need to develop both high-tech and conventional capabilities to participate in the RMA. In some of the articles excerpted above, improving the 'quality' – education level and technological sophistication – of PLA personnel as a prerequisite for embracing the RMA is discussed, and the Chinese were eager to learn more about U.S. training practices as soon as the post-Tiananmen restrictions on military contacts were eased. It would be a mistake to conclude from a review of the open-source material that this was all that China extracted from its monitoring of trends in the 1990s, however. Given the acknowledged asymmetry between the U.S. and Chinese militaries in this period, and the PLA's desire to use technology transfers to leapfrog, we should not expect to find blueprints for actions to redress the imbalance, at least not labeled as such.

Still, some PLA behavior following from the diagnosis of the situation in the 1990s could not be disguised. 'It wasn't the equal of ours, but it was impressive by any standard, and they did it in one year,' was the response in 1999 of an American visitor to a Chinese training center modeled on one that PLA visitors to the United States had toured in the late 1990s.[28] Other evidence of direct Chinese responses to the RMA includes steps taken to reduce the size of the force while increasing the degree requirements, especially in certain branches. As a result of these measures Chinese military writers can now boast that almost 80 percent of cadres in the Second Artillery have bachelor's degrees or higher, for instance.[29]

What we observe in this initial period, then, is an effort to gather information about the RMA and to prepare to respond through changes in PLA recruitment, training, and procurement. The emphasis on 'information', 'non-contact' and various other potentially benign-sounding kinds of warfare does not seem to indicate a bloodless interpretation of the RMA. Rather, PLA strategists believe that the U.S. achieved remarkable kinetic effects at range in the Gulf War and Kosovo through the manipulation of information available to the opposition and the ability to strike military targets from long distances. Finally, the PRC's approach to the *junshi geming* in this period seems to have involved continuing to reassure the United States about China's inferiority while concealing a nascent effort to confront the United States with an unfavorable balance.

Contemporary Chinese applications

In the current decade, as China has implemented its RMA strategy, we have witnessed a shift from a posture of reassurance toward greater willingness to demonstrate capabilities. During the 1990s, as China gathered data about the RMA and analyzed its effect on the balance of power, the PLA continued to pursue comprehensive modernization – from road-mobile missiles, upgraded nuclear forces, higher quality fighter aircraft and surface ships, and more stealthy submarines, to expanded and improved air defenses, mines, torpedoes, and the like – mostly through foreign purchases.[30] At the same time, consistent with Deng's injunction to 'bide time and hide capabilities', and the broader classical Chinese tradition of denying adversaries intelligence, these advances were made relatively quietly. China did not demonstrate its new weapons systems through tests aimed at other powers (with the important exception of the 1995–1996 Taiwan Strait crisis) and many of the new platforms were installed at remote, interior bases, with limited exercises that would have exposed them to public view. But in the last few years, beginning arguably with the 2007 anti-satellite test (ASAT), the world has seen more of China's modernized force structure. What is the logic behind the PLA's show of might through the ASAT test, increasingly prominent military activities in cyberspace, and development and testing of the DF-21D ASBM, along with an array of other ballistic and cruise missiles? Why are we increasingly reading about PLA exercises conducted in a complex electromagnetic environment?[31]

Again, in keeping with a strategic tradition that emphasizes secrecy, American readers do not have access to documents outlining the rationale for these gambits. But it is clear that from the perspective of the PLA strategists cited above, the capabilities that have been displayed or are rumored to have been acquired have tremendous disruptive potential. They are noncontact in that they would allow the PLA, at least in an initial strike, to inflict damage at range. Together with ground-based laser painting of objects in orbit and other incidents, the direct ascent kinetic-kill vehicle that China successfully shot into an aging weather satellite in January 2007 seems designed to send a message to the United States about the vulnerability of its reconnaissance and positioning assets in space. Cyber intrusions for the sake of espionage, denial of service, or sabotage similarly have the quality of giving China a way to signal the ability to disrupt American civilian and military operations. And the PLA's nascent ASBM is likely to offer China a potent first strike option against an American aircraft carrier.[32]

Does this mean that China has given up on 'hiding its capabilities and biding its time', confident that it can broadcast once-secret aspects of its defense posture? Considering the opacity that still surrounds the PRC's military budget, doctrine, and view of the dominant engagement in a potential conflict with the United States, the answer is clearly no. What has changed is that the PLA now uses uncertainty as a substitute for concealment, where hiding capabilities is no longer practical or desirable. Uncertainty reigns, not only about the budget but also about the true extent of China's cyber or network combat potential. Further, in the cyber domain, the PLA may be able to benefit from plausible deniability, complicating attribution for attacks. While there would likely be no mystery if a Chinese missile hit a U.S. satellite or aircraft carrier, the absence of any hints from PLA sources about the conditions under which they envision using such weapons is striking. Is the build-up all about Taiwan? If so, then why do the Chinese seem to have blue-water naval ambitions? The posing of such fundamental questions by foreign observers renders the partial revelations undertaken to date consistent with ancient Chinese counsel about maintaining information superiority.

In 2000, then-Senior Colonel (now Major General) Chen Bojiang spoke vaguely to an American interviewer about the connection between Chinese offensive capabilities and expectations for war:

> No enemy would 'let themselves so easily be involved in a protracted war with China,' though China might be defeated, because of the excessive cost of campaigning. Moreover, given overall Chinese strategy, 'It is also unallowable to have a protracted war. Under the conditions of new history, the main task of the country is to carry out the economic construction ... military actions must be [quickly accomplished in] scope and time.'

From this, Chen derives an emphasis on the offensive, according to Hawkins, who quotes him to the effect that 'attack as the main resort has an extraordinary importance on the high-tech battlefield'.[33]

What can be inferred is that the Chinese RMA vision is to acquire the capacity to inflict significant costs on an adversary, even a conventionally superior one, through a variety of means from targeting space assets and electro-magnetic pulse attacks to strikes on aircraft carriers and even civilian computer networks. Though new clues and hints have emerged, the full range of tools at the PLA's disposal remains enshrouded, encouraging the United States to err on the side of caution. Facing a potentially broad spectrum of Chinese destructive capabilities and lacking an understanding of the PLA's concepts of operation for using these weapons, U.S. decision makers might rule out challenging the PRC.

Conclusion: impact of the Chinese RMA on the military balance

Two questions remain to be addressed: is the account of Chinese strategy advanced here the most convincing interpretation? and, are Chinese military strategists correct in their estimate that the RMA alters the balance of power between China and the United States, neutralizing or supplanting U.S. conventional superiority?

The first question raises the issue of alternative hypotheses, including the argument that China is not seeking to challenge the United States or redress the balance but merely to claim its place as a great power. According to this hypothesis, the PLA's modernization is not distinctive but rather consistent with expectations derived from the record of other great powers' conduct. Given the range of great-power behavior observable in the twentieth century, it is difficult to know whether to be reassured by this line of reasoning. China could have modeled itself on post-war Germany or Japan and avoided the risk of antagonizing the United States. To be sure, that would have lowered the costs for a Taiwanese declaration of independence, but if Taiwan is the reason that the PRC has embarked on a massive buildup, then China's recidivism makes it a certain kind of rising power, and the acquisition of capabilities that indicate ambitions well beyond Taiwan remain to be explained.

A second alternative hypothesis would counter the details about the RMA above with evidence that the PLA is primarily focused on domestic security, with the bulk of forces still assigned to the army, attending to border control, disaster relief operations, and other conventional or internal missions. To this there is an easy response. One virtue of China's RMA strategy from Beijing's point of view is that it is consistent with a continuing investment in domestic stability forces. Domestic stability remains the Chinese Communist Party leadership's priority, and to date, the PLA has proven capable of serving it even as it has acquired specialized, high-tech capabilities and undergone organizational transformation in areas related to the fielding of these capabilities. For the PLA and its Party leadership, moreover, external and internal threats are linked, as described above, so that the ability to deter the United States may be seen as inseparable from the domestic security mission.

Turning to the question of the military balance, the Chinese strategists could err in thinking that the PLA's adoption of the RMA renders China capable of deterring, or if necessary, overcoming a challenge from the United States. Their fallibility may reveal itself in three ways. First is the possibility that they will misjudge what display of threat or force is sufficient to break American will and find themselves unprepared for resiliency in the face of what had been envisioned as a fait accompli. Second, to the extent that they have envisioned the application of a 'warfare engineering' or 'serialized' approach, they may fail to foresee the ways that a conflict could escalate – perhaps because the adversary turns out to possess and deploy hitherto unknown capabilities. Third, in the course of the PLA's selective revelation of new capabilities, the Chinese may find that they incite a response that they had not expected, either from a regional power or from the United States. This would disrupt whatever phased rollout had been planned and might even embroil the PLA in a conflict prematurely, prior to the full acquisition of assets necessary to defeat the enemy's will. One can imagine that there are debates within the PLA leadership ranks over how and when to test, and there is no guarantee that the right conclusions will always be reached.

Acknowledgments

The author thanks participants at the June 24–25, 2009 'Military Theory in the Post-Cold War Era' conference at the Norwegian Institute for Defence Studies in Oslo, Norway. She also thanks interlocutors at the Institute for Peace and Conflict Studies in Delhi on January 14, 2010, and Richard Bitzinger for his thoughtful review of her original *Journal of Strategic Studies* article, published by H-Net on April 11, 2011.

Notes

1 This evolution was predicted by at least one early American observer of the RMA:

> We live in a period of large scale, rapid technological and, very likely, social change. The pace of technological change is accelerating. We have not fully exploited and adjusted to developments in information and communication technologies; the next wave of change-producing developments is coming out of the biological and human sciences, which are likely to become significant sources of change in military operations and organizations.
>
> (Andrew W. Marshall, 'Forward' in *The Military–Technological Revolution: A Preliminary Assessment, 1991*, Center for Strategic and Budgetary Assessments, 2002)

2 Marshall, 'Forward', and in Andrew F. Krepinevich, *The Military–Technological Revolution: A Preliminary Assessment, 1991*, Center for Strategic and Budgetary Assessments, 2002.

3 Thomas Keaney and Eliot A. Cohen, *Gulf War Air Power Survey Summary Report*, Washington, D.C.: U.S. Government Printing Office, 1993, pp. 235–51; Stephen Biddle, *Military Power: Explaining Victory and Defeat in Modern Battle*, Princeton: Princeton UP, 2005.

4 Dima P. Adamsky, 'Through the Looking Glass: The Soviet Military–Technical Revolution and the American Revolution in Military Affairs' in *Journal of Strategic Studies*, 31:2 (April 2008) pp. 257–94; Adamsky, *The Culture of Military Innovation: The Impact of Cultural Factors on the Revolution in Military Affairs in Russia, the U.S., and Israel*, Stanford UP, 2010.

5 Xia Liping, 'China, U.S., Japan Strategic Relations: Striving for Win-win and Avoiding Security Dilemmas' in CPP20071109587001 Beijing *Shijie Jingji Yu Zhengzhi* (in Chinese), September 1, 2007.

6 Michael Pillsbury, *China Debates the Future Security Environment*, Washington, D.C.: National Defense University Press, 2000.

7 Wang Baocun, 'Military Reform in a Transformation Era' in FTS19980506000321 Beijing *Jiefangjun Bao* (in Chinese), April 21, 1998, p. 6.

8 Eliot A. Cohen, 'A Revolution in Warfare' in *Foreign Affairs*, 75:2, March/April 1996 , pp. 37–54.

9 Wang, 'Military Reform'.

10 Ibid.

11 Liu Kun, 'China: Torpedo Stealth' in FTS19960809000630 Beijing *Jianchuan Zhishi [Naval and Merchant Ships]* (in Chinese), August 9, 1996, p. 27.

12 Zhou Fangyin, 'The Effect of the Information Revolution on Military Affairs and Security' in CPP20010817000186 Beijing *Xiandai Guoji Guanxi* (in Chinese), August 1, 2001, pp. 28–32.

13 Wang, 'Military Reform'; Zhou, 'The Effect'; 'Watch Closely the Revolution of Military Technology in the New Era', interview with Zhu Guangya in FTS19951023000001 Beijing *Jiefangjun Bao* (in Chinese), October 23, 1995, p. 7; Peng Guangqian and Yao Youzhi (eds.), *The Science of Military Strategy*, English language edition, Beijing: Military Science Publishing House, 2005, pp. 431–3.

14 Robert Butler *et al.*, 'West Meets East: Chinese and Western Researchers Exchange Views on the Revolution in Military Affairs' in Historical Evaluation and Research Organization (HERO) Library. Online. Available at: www.herolibrary.org/p117.htm (accessed June 2009).

15 Butler *et al.*, 'West Meets East'.

16 PRC State Council, *China's National Defense in 2004*, Beijing: State Council Information Office, 2004.

17 Pillsbury, *China Debates.*
18 Jacqueline A. Newmyer, 'Oil, Arms, and Influence: The Indirect Strategy Behind Chinese Military Modernization' in *Orbis*, Spring 2009, pp. 205–19; Pillsbury, *China Debates.*
19 Note the contrast between this kind of materialist, arbitrary, and contingent political culture on the one hand and the modern Western liberal notion of rule by law, according to the consent of the governed, who are endowed with basic rights, including that of regular political participation on the other hand.
20 Sheryl WuDunn, 'China Says Soviets Erred Earlier in Picking Leader' in *New York Times*, September 8, 1991, section 1, p. 13.
21 *Selected Works of Deng Xiaoping, Vol. 1, (1938–1965)*, Beijing: Foreign Languages Press, 2006.
22 Ding Haiming and Sun Zhaoqiu, 'First Appearance of "Informatized Blue Army" in the Training Field' in CPP20050711000095 Beijing *Jiefangjun Bao (Internet Version)* (in Chinese), July 11, 2005, p. 2.
23 Shen Weiguang, 'Trends in the Development of World Warfare – Reducing Destructive Force' in *On the Chinese Revolution in Military Affairs*, Shen Weiguang (ed.), Beijing: New China Press, 2004 pp. 131–46, cited in Timothy L. Thomas, *Decoding the Virtual Dragon*; Xu Hezhen, 'Focus on Psychological War Under the Background of Larger Military Strategy in CPP20001211000122 Beijing *Zhongguo Junshi Kexue* (in Chinese), October 20, 2000, pp. 67–76; see also Xu Hezhen in CPP20011121000214 Beijing *Zhongguo Junshi Kexue* (in Chinese), September 30, 2001, pp. 94–100.
24 Ibid.
25 Fan Zhenjiang *et al.*, 'Military Theoretical Innovation Needed for Preparing for Information War, High-Tech War' in CPP20030121000089 Beijing *Jiefangjun Bao (Internet Version)* (in Chinese), January 21, 2003, p. 6.
26 Dai Qingmen, 'Discourse on Armed Forces Informationization Building and Information Warfare Building' in Shen Weiguang (ed.), *On the Chinese Revolution in Military Affairs*, Beijing: New China Press, 2004, pp. 39–47, cited in Timothy L. Thomas, *Decoding the Virtual Dragon.*
27 Dai Qingmin, 'On Seizing Information Supremacy' in CPP20030728000209 Beijing *Zhongguo Junshi Kexue* (in Chinese), Apr 20, 2003, pp. 9–17.
28 Charles F. Hawkins, 'The People's Liberation Army Looks to the Future' in *Joint Forces Quarterly*, 25, Summer 2000.
29 Andrew S. Erickson and David D. Yang, 'On the Verge of a Game-Changer' in *Proceedings*, 135:5, May 2009.
30 Richard Fisher, *China's Military Modernization: Building for Regional and Global Reach*, Westport, CT: Praeger Security International, 2008.
31 PRC State Council, *China's National Defense in 2006*, Beijing: State Council Information Office, 2006; Cao Qisheng and Huang Chao, 'Jinan Theatre Carries Out Special-Topic Study on Frequency Use' in *PLA Daily* (in English), June 12, 2009.
32 In a July 2011 English language *People's Daily* article, the PLA's Chief of the General Staff, General Cheng Bingde, acknowledged the ASBM program and described the capability as undergoing 'research, development, and testing'. In December 2010 Admiral Robert Willard, then commander of the U.S. Pacific Command, stated that the ASBM had reached 'initial operational capability'.
33 Hawkins, 'People's Liberation Army'.

7 Doctrinal innovation in a small state

Kjell Inge Bjerga and Torunn Laugen Haaland

Academic debate about military innovation and the IT-RMA has often focused on the experience of major powers, in particular on doctrinal development in the United States.[1] The evolution of military doctrine in small states has received less attention. Since military innovations rarely originate in such states, this is not surprising. It is nevertheless important to explore the particular characteristics of doctrinal development in states that are on the receiving end of the international trends in military thought and innovation discussed in this book.

In this chapter we will analyse the development of the Norwegian Doctrine for Joint Operations published by the Chief of Defence in 2007.[2] We will start by sketching the main characteristics of the Norwegian Armed Forces after the Cold War, including the early attempts to formulate a joint doctrine in the 1990s. Second, we will analyse the development of the 2007 doctrine, focusing, in chronological order, on three main topics: (1) inputs from the political and military leadership; (2) the organisation of the revision process; and (3) inputs from the military organisation. In conclusion, we will discuss to what extent the Norwegian experience is relevant for other small and large states inside and outside NATO.

Three key arguments will be presented. The first is that small countries have limited freedom to develop a separate way of thinking about the use of their military forces. Their dependence on allies and international institutions precludes any such attempts.

Second, small states also have constricted institutional capacity to develop their own military thinking and doctrines. Their research capacity is limited, and their military academies are relatively small with only rudimentary competence in many fields. Both arguments point in the same direction, namely that small countries' doctrines will be heavily influenced by the thinking taking place in major powers; only to a limited extent will they address particular national security concerns.

The third and final argument is that small countries' participation in large-scale multilateral operations is often partly or even primarily motivated by political symbolism. Force contribution signals political support for an international institution (NATO) or a common cause, but the forces are too small to make a real difference in the military operation as such. As a consequence the operational experiences of these forces differ from those gained by larger countries. The stakes are lower and they are unlikely to suffer military defeat in a decisive way. They may suffer losses, international embarrassment and domestic political turmoil, but they will not lose the war. They are saved by the fact that their influence upon events is much too limited. Their need to adjust and learn from operational experiences is therefore less pressing than that of a great power, and their military doctrines can develop fairly independently of operational experiences.

Doctrines and small states – two key concepts

Apart from the fact that the concept of doctrine itself means *that which is taught* (from the Latin *doctrina* meaning *teaching* or *instruction*), there is no single answer as to what doctrine is. On the contrary, quite different documents have been labelled doctrines. However, common to most is that they contain a state's defence and security policy guidelines, whether written or unwritten. For instance, various United States administrations have proclaimed the direction of their foreign and security policy in the form of doctrines; among the best known are the 1823 Monroe doctrine, the 1947 Truman doctrine, with the Weinberger and Powell doctrines from the 1980s and 1990s as the most recent additions. In Norway comparable foreign policy declarations have been formulated in Parliamentary Reports (White Papers) and in Propositions from the Ministry of Foreign Affairs and the Ministry of Defence (MoD). In 2004 the MoD also put forward its first strategic concept paper defining the country's defence and security policy in detail.[3]

The level under that in the doctrinal hierarchy consists of documents produced and published at the highest military level, stating how military force should be used in certain types of operations. In the Norwegian case, a joint doctrine addressing the entire spectrum of potential military operations was published for the first time in 2000. During the Cold War, the Armed Forces relied on NATO defence plans at this level.[4] In addition, each service and many of the branches within the services developed operational guidelines, including tactical directives.

Finally, at the lowest level of the doctrinal hierarchy a constant development and revision of operational concepts take place. Norwegian forces operating in Afghanistan today, and in the Balkans during the 1990s, have revised their procedures and operational concepts on the basis of their experiences in the field. Whether these experiences have made an impact on doctrine formulation at higher levels, is one question which will be explored here. That said, our main focus will be on the strategic military doctrines, which in Norway means the Joint Operational Doctrine. More precisely, we will look at the revision of the 2000 joint operational doctrine, which resulted in a revised version published in 2007. We will also sketch the development of doctrines and operational concepts prior to 2000. Norwegian defence and security policy will be briefly outlined as a background.

Just as the term doctrine covers several meanings, a country's ranking in an international hierarchy can be measured in several ways. Geography and the resource base, including territory and population, are traditional parameters. It has, however, been argued that today geographical size and the number of inhabitants count less than before and that other parameters have grown in importance.[5] In 2003, for instance, *Newsweek* rated Norway as ninth among the great powers of the world and Sweden seventh. The publication argued that the balance of power in the international system depended on historical context more than on square kilometres and population size. Decisive factors can be access to natural resources such as fish and oil, financial status, humanitarian engagement (great power 'in moral terms'), diplomatic skills and quality of life. The term 'smart states' has also been used to describe countries that are more influential and have a more prominent position in the international hierarchy than area and population alone would justify.

In this chapter, however, we restrict ourselves to military aspects, and here Norway remains a small state irrespective of the indicators applied. Although the Armed Forces have been modernised, they are limited in size. They also have restricted resources in terms of weaponry and equipment. All in all the Norwegian Armed Forces of today have little capacity to operate on their own, i.e. without the support of their allies.

Innovation after 1990: military experience, reform and doctrinal development

During the 20 years that have passed since the Cold War ended, the basic priorities in Norwegian security policy remain unchanged. NATO and the United States are the cornerstones of Norwegian security. For a small state, supporting the U.N. and upholding international law are key priorities. Relations with Russia, Norway's great neighbour in the north, have improved since the demise of the Soviet Union. Nevertheless, difference in size influences relations, as does the state of relations between Russia and the West. Finally, with Norwegian prosperity heavily reliant on offshore industries (oil, gas and fisheries), Norwegian security concerns have a distinct maritime dimension.

The important parameters of Norwegian security remain unaltered, but these parameters have themselves undergone fundamental changes over the last 20 years. As a result, the Norwegian Armed Forces have been transformed from a traditional mobilisation homeland defence force to a modern, flexible instrument of security policy. What does that mean in practical terms? The most striking aspect of the transformation concerns size. The fully mobilised Army has been reduced from 180,000 troops in

1990 to fewer than 10,000 in 2012.[6] For the Navy it has meant that all stationary anti-invasion capacities, most notably the fortification structure along the Norwegian coast, have been closed down. Many naval bases have also suffered the same fate. Furthermore, the number of vessels has been reduced from more than 90 to approximately 20. The Air Force infrastructure has also been downscaled and centralised. Its static operation concept has been overtaken by a more flexible and mobile concept based on smaller modules capable of being assembled and deployed to distant theatres at short notice (i.e. 'task force organisation').

Last, the organisation of the fourth service, the Home Guard, has also been centralised from more than 450 administrative units to the current 11 Home Guard Districts, and further reductions are due to be announced. Personnel numbers have been reduced from 83,000 to 40,000, of which fewer than 4,000 are to be trained on a regular basis to enable them to perform military tasks in case of crisis or war.[7] Parallel to these heavy reductions, training and equipment have improved, making a swift deployment nationally as well as internationally possible. These new abilities constitute an important qualitative aspect of the transformation process.

Apart from the changes in Norway's international surroundings, economy has been an important driving force behind the defence reform.[8] The gap between the financial means allocated and the number of tasks the military was expected to undertake widened throughout the 1990s. Budget deficits necessitated a much smaller and centralised structure to maintain operational capabilities.

New experiences from international operations were also important factors strengthening the reform drive.[9] During the Cold War Norway participated actively in U.N. peacekeeping operations, for instance on the border between Egypt and Israel (UNEF) in the 1950s and 1960s, in Congo (ONUC) in the 1960s and most notably, on the Israeli–Lebanese border (UNIFIL) from 1978 to 1998. Coming out of the Cold War, Norway took part in the first Gulf War and the failing U.N. operations in Bosnia and Somalia together with NATO allies. Throughout these years it remained a clear political priority not to send combat units to controversial operations abroad. These experiences were seen as irrelevant for homeland defence, and the operations had virtually no impact on the Armed Forces at home.

All this changed when NATO took over the U.N. operation in Bosnia in 1995. Political and military decision-makers recognised that political influence and military prestige was dependent on front line units rather than support functions, which were regarded quite unfairly as being less at risk and performing less demanding tasks. However, the Armed Forces soon discovered that its combat units had technical, organisational and psychological deficiencies, making rapid deployment into a conflict area impossible, a situation that led to painful reforms.[10] More robust units were allocated to the Stabilisation Force (SFOR) in Bosnia, Operation 'Allied Force' and Kosovo Force (KFOR) in Kosovo, and finally to Operation 'Enduring Freedom' and the International Security Assistance Force (ISAF) in Afghanistan. The general direction of developments has not changed since 1995: more robust contributions in operations, which allowed for more use of force.

Doctrinal development took place in parallel with both reform processes (restructuring at home and new contributions abroad). The first document resembling a joint operational doctrine was presented June 1995 under the title *Fundamental Views of the Chief of Defence on the Development and Use of Norwegian Forces in Times of Peace, Crisis and War.*[11] Although this document was not officially labelled a 'doctrine', it was the first written statement on how the military leadership in Norway would use the Armed Forces at a joint operational level. Major sections of the document were written by the Chief of Defence, and it was issued as an official directive from the Chief of Defence to the Norwegian Armed Forces; the top-down approach was unmistakeable.[12]

The document was rather short, a mere 24 pages, yet at the same time all-embracing. It included sections ranging from the political rationale of the Armed Forces, to broader strategic assessments, ending with Norway's relations with NATO and the structure of the Armed Forces. Most important, however, was that the Chief of Defence introduced manoeuvre warfare as the conceptual basis for Norwegian military operations, close to 15 years after this had been done in the U.S.[13] The introduction of manoeuvre warfare was related to a reform of the Norwegian Army called 'Division 2000', aiming at the creation of a modern army consisting of one full division capable of conducting high intensity warfare.[14] Besides the obvious inspiration from American and NATO doctrines, the Chief of Defence's conceptual basis appeared to be heavily influenced by Sir Basil Liddell Hart's writings on strategy, especially the 'indirect approach'.[15]

After the publication of the 1995 document, it seems that the idea of having a specific, written doctrine for the Norwegian Armed Forces gained support in wider defence circles. It became increasingly important for the Armed Forces to create a new conceptual basis for their activities, as defence planning in the traditional sense was abandoned in the aftermath of the Cold War. Furthermore, key decision-makers within the defence establishment were also inspired by the fact that doctrine played a more important role in NATO countries, for example in the U.K. and the Netherlands, than it had done previously.[16] From a Norwegian perspective it was regarded as important to develop national doctrines that promoted Norwegian particularities and concerns within the Alliance. Unlike, for example, the Danish Armed Forces, which opted for a continued reliance on NATO doctrines, the Norwegians emphasised concerns resulting from the asymmetric relationship with neighbouring Russia, and the vast sea areas under Norwegian jurisdiction.

In the year 2000 the Armed Forces' first official Joint Operational Doctrine (FFOD 2000) was published.[17] This was a considerably extended and augmented version of the 1995 document. Running to almost 400 pages, it consisted of two parts: Part A, *Fundamentals*, outlined the framework of military theory, as well as the political and geographical considerations governing the operational activities of the Norwegian Armed Forces. In Part B: *Operations*, the basic military functions 'central to the structural planning' and 'important concepts and factors associated with the planning and execution of operations' were outlined.[18] The basic idea underlying FFOD 2000 was *manoeuvre theory* and the principles, which were to apply in the execution of operations, were in line with those of *manoeuvre warfare*. This approach, albeit somewhat vaguely, aimed to (1) continue domestic reforms; (2) develop new contributions that could be deployed abroad; and (3) introduce a theoretical basis similar to the approach (already) adopted by the United States and most European NATO countries.

In retrospect the actual orientation of FFOD 2000 is easy to criticise. When it was published the Armed Forces were on the verge of a fundamental reform that reduced the size of the Army to a level at which large-scale manoeuvre operations were out of the question. Furthermore the doctrine's one-sided focus on manoeuvre warfare appeared to be of little relevance to the type of operations that the Norwegian Armed Forces were engaged in at the time (at home and abroad), operations which would only become increasingly important in the new millennium. Missions abroad consisted mainly of various types of peace support operations. At home the principal focus was on upholding national sovereignty, exercising authority and various forms of crisis management. In many of these operations the manoeuvre method, with its emphasis on speed, intensity and surprise, was of little relevance and did not contribute to the achievement of the overall political aims. Manoeuvre warfare was, however, tailor-made for the defence of North Norway against invasion from Russia, based on the belief that modest defence forces would be able through the use of smart moves, to resist and perhaps defeat, a stronger adversary. In other words the doctrine was adapted to a Norwegian context and the main national security concern, the only problem being that this national task was about to be abandoned as a result of the changing perceptions of the security environment. There appeared to be at least a two-year time-lag between security policy development and doctrinal development. A Naval Doctrine, Air Force Doctrine and Army Doctrine, all launched during the next couple of years, drew upon manoeuvre-based concepts in accordance with the 2000 Joint Doctrine.[19]

In 2007 the Chief of Defence published a revised joint operational doctrine. This doctrine differed from the previous ones in several important ways. Most importantly it did not present a normative guide on how to plan and conduct military operations. Instead, the doctrine sought to explain the new purpose and role of the military profession in the post-9/11 era. As stated in the introduction to the doctrine by the Chief of Defence: '[FFOD 07] can help the officer corps develop a common understanding, a shared way of thinking, a joint conceptual frame of reference and thus a foundation on which to build a common professional culture.'[20] The new doctrine was divided into six main chapters. After the introduction, in which the doctrine's ambitions and its place in the hierarchy of documents were explained, the different tasks of the Armed Forces were presented together with recent political, legal, ethical and technological developments influencing the execution of these tasks. Chapters three and four introduced several new concepts such as the effects-based and network-based approaches in addition to the traditional manoeuvre approach. The emphasis was on defining and explaining the meaning of these concepts in a rather theoretical manner. The topic of chapter five was joint operations, and the implications of jointness for the different services and branches were outlined. Finally, chapter six focused on the core values and ethics of the military profession. All in all the doctrine's main purpose seemed to be to present an updated description of the tasks of the Norwegian Armed Forces; secondly, to explain important aspects of the world that the Armed Forces were operating in; and last but not least, to introduce and define important new concepts which dominated allied and U.S. military thinking at the time. The doctrine listed several different types of operations, e.g. crisis management, stabilisation operations, asymmetric combat and regular combat operations. It outlined some principles for the different types but did not offer any operational guidelines for any of them. Instead, it seemed to signal that the concepts introduced would be applicable across the entire spectrum of operations listed.

In our discussion of the revision processes we will analyse how this change came about. However, we will first take a brief look at some theories seeking to explain the causes of such changes.

The sources of military doctrine

Theories about sources of military doctrine can be divided into three main categories. According to the first, military doctrine is a rational response to external threats.[21] Traditionally these threats have originated (geographically located) in other countries, but domestic threats should also be grouped here. Doctrines are designed to prescribe the most effective use of military force in relation to a given adversary and a defined political objective.

The second category sees military doctrines as the aggregated output formed primarily by organisational factors, e.g. routines, bureaucratic infighting, the interpretation of technological developments and civilian control mechanisms over the Armed Forces.[22] Organisational factors have an impact at the domestic level, but also at the international level in, for instance, NATO committees. Organisational explanations can be presented as an alternative to strategic explanations, or they can be presented as a filter through which strategic threats are interpreted and processed prior to organisational agreement on an appropriate response.[23]

A third category, which may be labelled 'constructivist', focuses on how doctrines are formed by perceptions of 'appropriate behaviour' within networks of professional officers educated at the same institutions, working in the same organisations, and sharing a distinct view of the world. A typical example of such explanations is the fact that military organisations prepare for major war, even though peace support operations and counter insurgency operations seem a more likely future scenario and a more effective way to meet external threats.[24] An explanation may be that major war doctrines are chosen because they offer prestige and recognition to the military organisation.[25]

None of the categories mentioned thus far are primarily concerned with the impact operational experiences have on the development of military doctrines. In fact, some of these theories go a long way in stating that doctrines are formed relatively detached from the military's operational experiences. However, in recent years the link between experiences and doctrinal change has received increasing attention, no doubt as a result of the difficulties Western militaries have encountered in Afghanistan and Iraq.[26] A focus on operational experiences introduces a bottom-up perspective into the analysis of doctrinal development. At the same time, the identification of lessons learned from these experiences is likely to be contested within the military organisation. Threat perceptions, organisational factors, and perceptions of appropriate behaviour will all impact on how the organisation interprets its own experiences. These processes are, however, not our main concern in this chapter. The question to be explored here is simply whether experience gathered in operations abroad had any influence upon Norwegian doctrine development in the process starting with the 2000 doctrine and ending with the 2007 doctrine.

Revising the Norwegian doctrine: political and military guidelines

Only two years after the publication of the Armed Forces Joint Operational Doctrine 2000, the first calls for revision were heard. This was a remarkably brief time span. Most likely, it reflected the striking discrepancies between the prevailing doctrine and the operations the Norwegian forces were increasingly involved in. However, the demands for revision did not originate in the fighting part of the organisation, perhaps because the 2000 doctrine was not significantly read or used in military operations. Another valid explanation, not contradicting the first one, is that the 2000 doctrine encapsulated military perceptions of how war should be fought. Even though it is difficult to trace the exact origins of the revision initiatives, it is likely that they came from the only part of the organisation using the doctrine in its daily work, namely the military education institutions. The Norwegian Command and Staff College (which in 2005 became part of the Norwegian Defence University College) and the three war colleges (Army, Navy and Air Force) had been the primary users of the 2000 doctrine. The Staff College was also responsible for doctrinal development at this level in Norway.[27]

When calls for revision began to be made in late 2002 and early 2003, the matter caught the attention of the Chief of Defence. In May 2003 he ordered his Joint Staff at the Norwegian High Command to prepare to revise the doctrine. A few months later the Norwegian central defence organization was thoroughly restructured. The Chief of Defence and the parts of the High Command involved in strategic planning, including policy and doctrinal development, were integrated with the MoD.[28] The merger of top level military and political institutions brought the matter of military doctrine closer to the Defence Minister. Consequently the political leadership became more involved in the formulation of doctrine than previously. Whereas the Ministry had been scarcely involved in the writing of the 2000 manoeuvre doctrine, it became heavily involved in the revision process from August 2003 onwards. The main explanation for this involvement was both the new organisation and the fact that the Ministry was about to publish its strategic concept for the Norwegian Armed Forces.[29] According to the Ministry a revised military doctrine at this level had to be in line with the new concept. Thus, in 2004, both the Defence Ministry and the Chief of Defence provided detailed guidelines for the contents of the new doctrine. As a consequence the revision process was organised from the top-down, rather than bottom-up method that characterised the initial phase.

Political guidelines

The Ministry of Defence's guidelines were expressed in the Long-Term Defence Policy Plan (White Paper) presented to Parliament in March 2004.[30] In short, the political objectives were twofold: to import new phrases and concepts already introduced by the Bush/Rumsfeld administration, and to signal the continued priority of national defence tasks. Whether they were compatible was a question the Ministry did not address.

An illustration of the imported American jargon was the Ministry's definition of doctrinal development: new defence technology, lessons learned, experimentation and concept development (CDE) were 'four elements' that jointly constituted a so-called 'wheel of transformation'.[31] The Ministry also stressed that doctrine had to reflect new technological trends, including innovations such as precision-guided munitions and information technology, and concepts such as asymmetrical warfare and network-centric defence.

At the same time, the Ministry emphasised that the revised doctrine had to reflect Norway's specific geographical, political and cultural circumstances, and that the doctrine should give priority to the Armed Forces' national tasks, including upholding national sovereignty, the exercise of authority and national crisis management. All in all the Ministry of Defence seemed to view the revised doctrine not as a comprehensive guide for military action, but as a communication channel suited to signalling the government's political priorities to allies, a domestic audience and the Armed Forces. The message was that Norway would continue the transformation of the Armed Forces towards a small high quality expeditionary force. Moreover, the government was eager to show the electorate that it was taking national security concerns seriously.

Military guidelines

The Chief of Defence's guidelines for the revision process were quite similar to the political ones. Like the politicians, he wanted to use the doctrine to enhance the defence reform and to communicate a message to a broad audience. In fact the Chief of Defence explicitly stated that the revised doctrine should serve a different purpose from the previous one.[32] First of all, the doctrine should support the ongoing transformation of the force structure, and was thus intended as a further measure to achieve the long-term goals formulated in the quadrennial defence reviews.[33]

Second, arguing that the use of military force was more controversial than before, and that the Norwegian Armed Forces were more in the public eye after the turn of the millennium because of its foreign deployments, the Chief of Defence advocated that the doctrine should address a broader audience than the narrow military community. This was a novel approach compared both to the 1995 and the 2000 doctrines, both intended solely for the Armed Forces.[34] The revised version was intended as an explanation of the new role that the Armed Forces had developed during the 1990s, with small semi-professional forces ready to deploy together with allied forces in complex international conflicts. That this explanation would endow the document with a degree of legitimacy was an added bonus.

Besides (1) enhancing defence reform, and (2) legitimising Norwegian deployments abroad, the doctrine should also (3) provide broad guidelines for the application of Norwegian Armed Forces in national and international operations, (4) elevate the level of competence among military personnel regarding the use of military force in the new kind of international operations, and finally (5) the revised doctrine should be a Norwegian input to allied doctrinal development.[35] The ambition was nothing less than to be at the forefront of allied doctrinal development, and thus influence allied doctrine. This last point was, of course, unrealistic for a small state like Norway, but it represents a typical 'Norwegian approach' to international institutions.[36]

Regarding the *contents* of the new doctrine, the Chief of Defence appeared to be strongly influenced by allied conceptual and doctrinal trends.[37] He stated that the doctrine should give priority to joint operations and to the technology these operations required. The influence of NATO's Allied Command Transformation (ACT), and thus from the American documents *Joint Vision 2010* and *Joint Vision 2020* (both based on the idea of a Revolution in Military Affairs) was strong.[38] The doctrine had to address key transformation concepts such as Network Centric Warfare and Effects Based Operations and explain their relevance to a Norwegian context. Both concepts were American imports, specifically from the U.S. Air Force. Their impact could be seen in allied doctrines through the activities within the Allied Command Transformation at this time.[39] Other important aspects of the new doctrine were military ethics and international law. The doctrine also defined key terms and introduced a common language for the military community.

Concerning *process*, the Chief of Defence encouraged a bottom-up approach: 'It is crucial that the editorial board listens to personnel and units with experience from current operations'.[40] He authorised the Board to establish networks including relevant persons and units, and to conduct interviews, arrange workshops, etc. The new doctrine was to be rooted in all parts of the Armed Forces, and this was to be achieved through close dialogue and coordination with all parts of the organisation. Nevertheless, his own detailed guidelines were clearly at odds with a real bottom-up review. Involving all parts of the organisation seemed above all to be intended as a 'sell in' of the new doctrine in order to prevent organisational resistance at a later stage.

All in all the prime concern in great power military doctrines, i.e. establishing guidelines for the use of military force and how to conduct military operations, was not as urgently felt by the Norwegian decision-makers. Their ambitions for the Norwegian doctrine were more related to public relations and current defence reform.[41] This was partly a result of organisational factors. Merging the political and military leadership inevitably made the doctrine more of a political document, removing it from the traditional military functions of planning and conducting operations. At the same time the strong emphasis placed on transformation and theories in vogue in the U.S. administration seemed conditioned by perceptions of 'appropriate behaviour', i.e. adapting Norwegian thinking about military operations to correspond with American and allied thinking in this field. By applying terms and expressions borrowed from the great military powers, the Chief of Defence and the Defence Ministry sought prestige and recognition for the Norwegian military organisation at home and abroad. However, the unabridged adoption of U.S. and allied transformation concepts may also be regarded as rational. In 2006 Norway scaled down its ambition to be able to resist an invasion on its own for a limited period of time. Territorial defence, which had been the

Norwegian Army's main task since 1628, was transferred from the Army to the Home Guard.[42]

In so doing, Norway became even more dependent on allied support than ever before. Being able to operate with allies from day one became *correspondingly important*. If Norway was to develop its own defence concepts, dressed in a different language to that of NATO, cooperation and interoperability would most likely suffer.[43] A rational analysis linking doctrinal development to specific threats confronting Norway is, however, hard to detect in the viewpoints expressed by the political and military leadership.

Organising the revision process: complexity, crisis, collapse

After the political and military directives had been formulated, the revision process continued inside the military bureaucracy without interference from the top layer.[44] In this section we will analyse how the organisation of the revision process left its imprints on the 2007 doctrine. The Norwegian Command and Staff College was to be responsible for the final text, preferably in close dialogue with relevant parts of the Armed Forces. The Chief of Defence had stressed the importance of this dialogue, for two main reasons. First, since the Command and Staff College had only limited experience from current operations, it was crucial for them to get into close contact with those parts of the Armed Forces that had. Second, the dialogue was seen as a *sine qua non* for endowing it with the requisite legitimacy within the Armed Forces. In order to achieve that, the following organisation was established.

At the top level, the Armed Forces Concept and Doctrine Forum was set up in 2003. Its purpose was to bring together different levels and institutions of the defence organisation, and thereby establish a common understanding of the revision process and the doctrine. This was regarded as crucial for the legitimacy of the doctrine. The forum was headed by the Defence Ministry, and was intended to offer advice and guidelines in the write-up of the doctrine.[45] For practical reasons commenting on drafts was delegated to a *Secretariate* consisting of mid-level military and civilian personnel from the Defence Ministry.

The actual writing process was left to the *Editorial Board* composed of five officers employed at the Command and Staff College. Between them these five represented the Air Force and the Army, while the Navy was unintentionally left out.[46] They were mid-level officers (Major/Lieutenant–Colonel), with a traditional military education but a very limited experience from operations. These officers were relieved from other tasks and fully devoted to the revision process. In addition, several working committees were specifically allocated to areas like: defence technology, command and control, Effects Based Operations (EBO) and Network Based Defence (NBD), i.e. the Norwegian approach to Network

Centric Warfare.[47] Finally, board members spent time in the U.K. and other allied countries, as well as in NATO institutions, to collect first-hand information about international trends pertaining to conceptual and doctrinal development.

As work progressed, the *dialogue* between the Forum and the Editorial Board became the focal point of the writing process. However, a widening gap between the members of the Forum and the Editorial Board soon made itself felt. The complex organisation of the Forum made a functional division of responsibilities increasingly difficult. Perhaps more importantly, the agendas of the various bodies of the Forum left little time for doctrinal matters. Although they met on regular basis, it was not frequent enough to instruct or guide the writing process in any detail. A growing asymmetry developed between the bodies of the Forum and the Editorial Board with regard to how devoted they were to the doctrinal revision, how updated they were on specific doctrinal topics, and how familiar they were with international trends and allied doctrinal development. Furthermore, it turned out to be very challenging for the members of the Forum to read and comment upon draft texts written by the Editorial Board while at the same time retaining their full-time senior positions in the defence sector. Thus, for the simple reason of capacity, the five mid-level officers on the Board became far more influential than the other members.

The actual composition of the text started at the Staff College in the Autumn of 2004. The ad hoc committees put forward supporting material and ideas in note form, as well as drafting sections of the text. Throughout the period from Autumn 2004 to Spring 2005 several rounds of consultative meetings and seminars were held. These had a clear bottom-up profile with a broad remit, and were intended as a means of analysing and discussing the contents of the doctrine on a continuous basis. Anchoring and participation were the key words. However well organised it may have been, by May 2005 the process came to an abrupt halt for three main reasons.

First, the text drafted by the Editorial Board was not well received by the Secretariat, which claimed that major revisions were required if the new doctrine was to be accepted by the military leadership as well the different parts of the military organisation. With regard to contents, disagreement emerged over: (1) whether the doctrine should include strategy and security policy, and (2) the priority given to RMA-inspired ideas on warfare. The Secretariat claimed that the Editorial Board had promoted these ideas too eagerly, and the resulting distance to allied publications and doctrines was too large. The Editorial Board responded by saying that it was adhering to the initial guidelines drawn up by the political and military leadership, and that, ultimately, competence in this field rested with them. From their point of view, it was also important to uphold the ambition of being at the forefront of allied doctrinal development.

Secondly, in April 2005 the recently appointed Chief of Defence, General Sverre Diesen, issued new guidelines for the doctrine developers. Ruling in favour of the Secretariat, Diesen simply ruled that all references to defence and security policy should be removed and left to a recent publication by the Defence Ministry entitled Strategic Concept for the Norwegian Armed Forces (Relevant Force), in which defence and security policy directions were spelled out in detail. Furthermore, the new Chief of Defence rejected the idea of the doctrine outlining procedures for both the planning and conduct of operations. Since procedures were continuously being revised, a doctrine should be restricted to drawing up general and timeless principles of force application. Finally, he stated that the doctrine had to be brief and to the point in order to be read. According to him, 150 pages sufficed to outline fundamental ideas about the use of Norwegian military force. As a result large sections of the text produced had to be scrapped.[48]

A third and final reason for the process coming to an end was that the organisation established for the writing process proved to be too complex. It was difficult, if not impossible, to combine the Forum's recommendations, including those coming from its numerous bodies, ad hoc working committees, not to mention inputs from the lower levels of the organisation, into a single coherent doctrine. Furthermore, maintaining a dynamic writing process within such a huge and formalistic structure proved close to impossible. The Editorial Board was forced to work with drafts that were being continuously circulated and revised. Not surprisingly it failed to come up with a final, comprehensive, version.

During the summer of 2005 the Board disintegrated. In Autumn 2005 a new *Writing Committee* was appointed by the Head of the Command and Staff College, this time consisting of only two persons: a civilian and an officer, both academics but neither with any operational experience.[49] The idea was to 'close the process', i.e. to let the new committee synthesise all the draft texts produced so far (at this stage they ran to more than 500 pages), in order to come up with a final, complete and comprehensive draft for a revised doctrine. This draft was then to be accepted and anchored both upwards and downwards in the organisation through a process of hearings. Hence it became important for the Writing Committee to produce a comprehensive draft doctrine in line with expectations from both the top and the bottom of the organisation. An unofficial, much smaller and more flexible *Shadow Council* was set up. It consisted of one representative from the Ministry of Defence, one from the Chief of Defence, and two representatives from the Command and Staff College. As a result the actual writing of the doctrine was undertaken by a small group of persons without any operational experience, situated inside a military education institution and with the Defence Ministry as their closest counterpart.

The final stage: inputs from the military organisation

In the autumn of 2005 work on the doctrine intensified. The following year a comprehensive draft version was circulated for comments within the military organisation, the Ministry of Defence and the Defence Research Establishment. The involvement of the military organisation at this stage was substantial. Some 28 different units came forward with more than 1,400 detailed revision proposals.[50] On the one hand the all-embracing process generated valuable inputs; on the other hand many were too biased to be rendered useful.

First, most inputs came from those parts of the military responsible for administration, education, and support functions in military operations. In particular the military academies and the single service staffs sent in quite detailed comments. Specific professional communities within the military, such as the ones dealing with intelligence, logistics, and information systems, also voiced strong views on the doctrine's contents. At the same time, eliciting responses from those responsible for the planning and conduct of operations turned out to be unexpectedly difficult. The Joint Operational Headquarters, which would be a primary user of the joint doctrine, failed to respond. Most likely the activities at the Headquarters, including the planning and execution of operations abroad and at home, were directed by guidelines and directives in the chain of command and not by the doctrine.[51] Moreover, the operational community at the Headquarters was 'practically minded'. Digesting and reacting to a doctrine in which cultural change and theoretical approaches were central may all too easily have been dismissed as a waste of time. Furthermore, there were no structures or mechanisms that connected the existing Centre for Military Experience, or any other of the military units working on operational lessons, to the process of revision. As a consequence these communities displayed at best only a limited interest in the doctrine.

As a result the Writing Committee received few inputs from sections of the military connected to experiences or lessons learned from former or ongoing operations. Instead most reflected the institutional interests of different services, branches and educational units. Among the more prominent were comments from the Intelligence Service, the Logistics Organisation and the Medical Service. All of these claimed that insufficient attention had been devoted to their activities. The logisticians pointed out that 'readers of the doctrine must not be misled to believe that the need for logistical readiness is any less today than earlier'.[52] The Intelligence Service took exception to the fact that the Writing Committee had described Effects Based Operations and Network Based Defence without explaining the importance of intelligence,

> which is a serious weakness because intelligence is a very central function

> [...]
> [W]e might get the impression that the need for intelligence is replaced by these concepts while the case is that these very concepts increase many times the need for intelligence.[53]

The Medical Service was left unmentioned in the first draft circulated for comments, which of course triggered a strong reaction: 'The Medical Service is puzzled by the fact that Medical Service is not mentioned at all in the new doctrine.'[54] The Medical Service then suggested a new chapter devoted to this topic alone.

Not surprisingly the Air Force Academy and the Air Force Staff applauded the emphasis placed on effects-based approaches in the new doctrine.[55] The Air Force Staff encouraged emphasis on this specific approach, arguing that the doctrine should further elaborate on the advantages of the concept of Effects Based Operations and highlight the linkage between this concept and the concept of Network Based Defence.[56] The Air Force also argued that the examples used in the doctrinal draft were 'dominated by army operations' and that it was important to underline the importance of air power and the Air Force. Not least, they argued, since the doctrine described day-to-day operations in support of Norwegian sovereignty and authority, in which the air defence surveillance system in combination with combat aircraft at a high state of readiness played important roles.[57]

The Army's transformation and doctrine command (TRADOC) took the opposite view and applauded the high priority given to the manoeuvre approach and manoeuvre warfare. It nevertheless expressed scepticism about the description of effects-based approaches and the concept of Network Based Defence. TRADOC found both to be 'very abstract'.[58] Without causing any surprise at the receiving end, the Army also advocated that the doctrine should put even greater emphasis on the importance of joint operations, and in more detail describe *how* the tasks of the Armed Forces could be solved through joint operations.[59]

The Naval War College deviated little from the Army line. The College supported the use of manoeuvre approach and manoeuvre warfare, but remained suspicious of the introduction of the effects-based approach, and the concepts of Effects Based Operations and Network Based Defence.[60] However, the main input from the Navy was dominated by the strong position John Boyd's (1927–1997) theory of decision-making in combat (the so called 'OODA-loop') enjoyed at the Naval War College in Bergen.[61] The War College argued that Boyd's theory of decision-making should be included in the doctrine and that theories on leadership in general had to be featured more prominently.

Other comments from below were at odds with the original political and military directives. For instance, the Department of Joint Operations at the National Defence University College argued that the doctrine was far too theoretical, and that a doctrine should provide specific guidelines for joint operations similar to NATO doctrines.[62] Some of the officers at the Department of Joint Operations also questioned the very idea of having a national doctrine. As they saw it, future deployments of Norwegian Armed Forces would take place within an allied context, i.e. in operations guided by NATO doctrines.

After the broad remit in 2006, the Writing Committee finalised the doctrine in Spring 2007. Many of the inputs from the services and specific professional communities were included in the doctrine. For example, John Boyd's OODA-loop was given a prominent place, the need for logistical readiness was underlined, as well as the need for intelligence in all kinds of operations, and the Medical Service was described properly. From the perspective of the Writing Committee, listening to the organisation was required if the legitimacy of the doctrine was to assured. However, this process did not link the doctrine to the operational experiences of the Norwegian Armed Forces, at least not in any direct way. Since inputs remained more concerned about defending one's own turf against intruders rather than discussing principles for the conduct of current and recent military operations, the final doctrine was completely detached from the operational level of the organisation.

Conclusion: doctrinal innovation detached from military operations

After having been approved by the Defence Ministry and the Chief of Defence, the doctrine was published in June 2007, and by 2010 more than 25,000 copies in Norwegian and English had been distributed. The discussion presented here has shown that from 1995 onwards, development of military doctrine in Norway has gradually been separated from operational experiences and the sections of the Armed Forces that plan and conduct operations. During the last 15 years doctrinal development has primarily been attached to academic and educational institutions within the Armed Forces, while the operational sections of the forces, including the Joint Operational Headquarters, have shown only limited interest in doctrine. Instead these sections of the forces remain reliant on directives and guidelines that are part of the official chain of command, which the doctrine is not. This separation between operational activity and the joint doctrine is reflected in the contents of the Norwegian doctrine. The theoretical orientation is strong.

While the Chief of Defence himself was heavily involved in the writing of the 1995 doctrine, and this doctrine was closely linked to a specific force structure, both the 2000 and 2007 doctrines were authored by academics at the Command and Staff College. They underlined that the ideas presented were universal and therefore neither dependent on any specific structure of the Armed Forces nor any known threats to Norwegian security. The purpose of the current Norwegian doctrine is not to enhance operational performance in relation to specific security risks. More correctly, it can be described as a tool for internal and external communication.

To what extent are these findings relevant for other states inside and outside NATO? The analysis of the revision process presents some general insights. Most striking with regard to organisational factors remains the fact that those who have the time and the ability to do the actual writing are also in a position to exert great influence on contents. Country size plays no role here, one might add. This is perhaps the case in most countries, irrespective of size. In Norway, the Chief of Defence and the Defence Ministry provided broad guidelines in the beginning, after that they played a somewhat detached role. The academic communities involved in the actual writing of doctrine were left to do much of the thinking and writing on their own. They were far more influenced by international trends in military theory, trends originating in the great powers, than by experiences gathered from concurrent Norwegian military engagements. At the same time the operational part of the organisation showed scant interest in bringing their experiences into doctrinal development. The procedures created for broad organisational input were used to promote the interests of specific branches and to profess their continued relevance to the defence structure. Inputs linking doctrine to ongoing operations in Afghanistan or in the Barents Sea were close to non-existent. Such bureaucratic turf battles are of course neither new nor restricted to a Norwegian context, but can be found in all military organisations inside and outside NATO.

The revision process shows that importing NATO and U.S. language and concepts was given an overriding priority at the expense of developing a doctrine tailor-made for the challenges facing the Norwegian Armed Forces. On the one hand the final product is a doctrine in harmony with American and NATO doctrines. On the other hand the relevance in a Norwegian context of the approaches and operational methods outlined in both the 2000 and the 2007 doctrine is questionable. Hew Strachan pinpointed this in his review of the 2007 doctrine:

> All three of the approaches on which the doctrines foundation is based were developed specifically in the context of war, not least because its origins are to be traced to the United States, the most committed of the NATO powers to the centrality of war-fighting.[63]

War-fighting is of course central to the Norwegian forces as well, but concepts developed for the use of force of a great power may not be directly applicable for a small state like Norway. Furthermore, the approaches described in the doctrine may also not be applicable for another primary concern of the Norwegian Armed Forces, which is its peacetime activity. To what extent these activities, whether upholding sovereignty at home through territorial surveillance or participation in international peace support operations, can best be pursued through a network-centric and effects-based approach, was left out of the revision process. Having a doctrine reflecting international trends remained more important than addressing the use of force in ongoing operations.

However, being a committed member of NATO and nurturing a special relationship with both the U.K. and the U.S., makes it difficult for Norway to introduce totally different approaches and methods in its military doctrines. One aspect is perceptions of 'appropriate behaviour' within the Alliance, while another is the need for interoperability in an allied context. The importance of interoperability has increased because today's Norwegian Armed Forces have little autonomous operational capacity. Accordingly they only have limited relevance when operating outside a NATO context. This is a result of a broader development since the end of the Cold War whereby the Armed Forces of most small- and medium-sized European powers have reduced their ability to operate on their own, instead increasing their dependence on NATO and the United States.

This brings us to our last point with regard to doctrinal innovation in the particular case of small states. Being a small country implies that both ambitions and stakes are lower when military force is applied in large-scale multilateral operations. It is the great powers that run these operations and it is the reputation and credibility of their Armed Forces that is linked to their outcome. Because of this simple fact it is possible for small countries to write doctrines that have purposes other than guiding their current and future military operations. As we have seen in our discussion, both the political and military leadership had far-reaching ambitions for the 2007 doctrine beyond improving the operational performance of Norwegian forces deployed in current international operations. The leadership wanted a doctrine that (1) contributed to the transformation of the Norwegian Armed Forces in line with the U.S. and NATO military transformation at that time; (2) contributed to a so-called 'cultural change' in the mindset of Norwegian officers; (3) showed to the Norwegian public and the political establishment that the Norwegian Armed Forces had left the Cold War paradigm behind and entered a new era; and (4) showed to the international military community, above all the U.S. and the U.K., that Norway was a loyal ally which wanted to import military theory developed in the United States for a military several hundred times the size of the Norwegian military, and for a state with clear ambitions of global power projection. Whether these theories were suitable for a mini-force such as the Norwegian one could safely be left unanswered since the doctrine was detached from the question of operational effectiveness.

Notes

1 This bias towards the United States has a natural explanation, i.e. that much doctrinal innovation is American. Most notable in recent years has been the

development of the IT-RMA. See Stephen Peter Rosen, chapter 4: 'The Impact of the Office of Net Assessment on the American Military in the matter of The Revolution in Military Affairs'; Dima Adamsky, 'American Strategic Culture and the U.S. Revolution in Military Affairs' in *Defence and Security Studies*, 2, 2008, pp. 13–18. For a broader analysis of the impact of cultural factors on the course of military innovations, see: Dima Adamsky, *The Culture of Military Innovation. The Impact of Cultural Factors on the Revolution in Military Affairs in Russia, the U.S. and Israel*, Bloomington: Stanford University Press, 2010.

2 *Norwegian Armed Forces Joint Operational Doctrine*, Oslo: The Defence Staff, 2007. The Joint Operational Doctrine 2007 was published both in Norwegian and English. The doctrines of 1995 and 2000 were both published in Norwegian only.

3 *Strength and Relevance. Strategic Concept for the Norwegian Armed Forces*, Oslo: The Ministry of Defence, 2004. Five years later this concept was revised and published again, confirming the determination of the Norwegian Government to maintain a written security policy doctrine on this level: *Capable Force. Strategic Concept for the Norwegian Armed Forces*, Oslo: The Ministry of Defence, 2009.

4 In our context it is important to distinguish between *defence plans* and *doctrines*. Defence plans traditionally outline the forces required for the defence of a given area. During the Cold War NATO, and Norway as part of the Alliance, had defence plans as its focal point. 'Doctrines' in the present understanding of the word hardly existed at the strategic and the operational levels, and if so then only in a peripheral role. The Norwegian Defence Staff issued a *National Defence Plan* in 1949, but when Norway entered the Alliance's joint command and force structure in 1951 this national plan was replaced by NATO's defence plans. Until 1994, NATO's Northern Command (AFNORTH) maintained (i.e. wrote, revised and updated) the overall defence plan for the Northern Flank (including Norway). When AFNORTH was closed down in 1994 a vacuum developed regarding defence plans. However, in 1995 and 2000 Norway issued *doctrines*, which to some extent replaced the defence plans, even though the doctrines were very general and refrained from specifying how the defence of a specific area was to be conducted. Moreover, in December 2000 the Chief of Defence issued a *Strategic Defence Plan*, which was replaced by a *Strategic Directive for Operations* in April 2003 (updated and reissued December 2009). Both were classified. Compared to previous defence plans the latter 'plans' were detached from territorial defence. Instead they described procedures for the planning and conduct of operations at home and abroad (e.g. the organisation, deployment and withdrawal of a Norwegian unit in an international operation). See: Kjell Inge Bjerga and Kjetil Skogrand, 'Securing small-state interests: Norway in NATO' in Mastny, Vojtech, Sven G. Holtsmark and Andreas Wenger (eds), *War plans and alliances in the Cold War*, London: Routledge, 2006, pp. 218–40.

5 Andras Løvold, 'Småstatsproblematikken i internasjonal politikk' [Small state challenges in international relations] in *Internasjonal politikk* No. 1, 2004, pp. 7–13.

6 Kjell Inge Bjerga and Gullow Gjeseth, 'Heimevernet og Hæren: Landforsvaret stykkevis og delt – eller helt' [The Home Guard and the Army: a Divided or Unified Territorial Defence?]' in *Oslo Files on Defence and Security*, 2, 2010, p. 9.

7 Ibid., 12–13.

8 St. prp. nr. 45 (2000–2001), *Omleggingen av Forsvaret i perioden 2002–2005* [White Paper No 45: The transformation of the Armed Forces 2002–2005], pp. 11–12.

9 For more details, see: Torunn Laugen Haaland, *Small Forces with a Global Outreach. Role Perceptions in the Norwegian Armed Forces after the Cold War*, Oslo: Unipub, 2008.

10 Ibid.
11 *Forsvarssjefens grunnsyn for utvikling og bruk av norske militære styrker i fred, krise og krig* [The Chief of Defence's Fundamental views on the development and use of Norwegian forces in times of peace, crisis and war], Oslo: Forsvarets overkommando, 1995.
12 General Arne Solli (CHOD Norway October 1994–April 1999), interview 12 December 2009.
13 *U.S. Army Field Manual 100–5: AirLand Battle Doctrine*, HQ, Dept of the Army, 1982; *U.S. Army Field Manual 100–5: AirLand Operations*, HQ, Dept of the Army, 1993. Martin van Creveld, *The Transformation of War*, New York: Free Press, 1991; Robert L. Leonhard, foreword by Colonel J.R. McDonough, *Fighting by Minutes. Time and Art of War*, Westport Connecticut: Praeger, 1994.
14 Gullow Gjeseth, *Hæren i omveltning 1990–2005* [Transforming the Norwegian Army 1990–2005], Bergen: Vigmostad & Bjorke, 2008, pp. 73–83.
15 Basil Liddell Hart, *Strategy*, New York, NY: Meridian, 1991, 2nd revised edn.); Basil Liddell Hart, 'Strategy: the indirect approach' in Thomas G. Mahnken and Joseph Maiolo (eds), *Strategic Studies – a Reader*, Oxon/New York: Routledge, 2008. The writings of Liddell Hart were taught at the Norwegian Staff College at this time, and the Norwegian Defence Research Establishment used Liddell Hart's theories and other theories on manoeuvre warfare in their planning during the 1980s and 1990s. See: Torgeir E. Sæveraas and Kjetil Henriksen, 'Et militært universalmiddel? Amerikansk "Maneuver Warfare" og norsk doktrineutvikling' [A Military Wonder Medicine? American "Maneuver Warfare" and Doctrine Development in Norway] in *Oslo Files on Defence and Security*, 1 2007, pp. 109–24.
16 Marcel de Haas, 'Dutch Defence Doctrine', *The Officer Magazine*, 18:1, January/February 2006; *Defence Doctrine*, Haag: Ministry of Defence, 2005; *British Defence Doctrine*, London: U.K. Ministry of Defence, 2001.
17 *Forsvarets fellesoperative doktrine Del A: Grunnlag, del B: Operasjoner* [The Armed Forces' Joint Operational Doctrine, part A: Fundamentals, part B: Operations], Oslo: Forsvarets overkommando, 2000.
18 *Forsvarets fellesoperative doktrine Del B: Operasjoner* [The Armed Forces' Joint Operational Doctrine, part B: Operations], Oslo: Forsvarets overkommando, 2000.
19 *Forsvarets doktrine for luftoperasjoner* [The Armed Forces' Air Operations Doctrine], Oslo: Forsvarets overkommando, 2002; *Forsvarets doktrine for sjøoperasjoner* [Armed Forces' Naval Operations Doctrine], Oslo: Forsvarets overkommando, 2002; *Forsvarets doktrine for landoperasjoner* [The Armed Forces Land Operations Doctrine], Oslo: Forsvarets overkommando, 2004. Unlike the Joint Operational Doctrine, the three service doctrines have not been revised and updated.
20 *Norwegian Armed Forces Joint Operational Doctrine*, Oslo: Forsvarsstaben, 2007, p. 3.
21 John Gooch, 'Military Doctrine and Military History' in John Gooch (ed.), *The Origins of Contemporary Doctrine*, Strategic and Combat Studies Institute, 30, September 1997, pp. 5–6; J.F.C. Fuller, *A Military History of the Western World*, Vol. 3, New York: DaCapo Press, 1956.
22 Barry Posen, *The Sources of Military Doctrine*, Itacha: Cornell University Press, 1984, pp. 44–5; Jack Snyder, *The Ideology of the Offensive: Military Decision-Making and the Disaster of 1914*, Itacha: Cornell University Press, 1984, p. 104, p. 213.
23 Theo Farrell and Terry Terriff (eds.), *The Sources of Military Change. Culture, Politics, Technology*, Boulder CO: Lynne Rienner, 2002.
24 Olof Kronvall, 'Finally Eating Soup with a Knife? A Historical Perspective on the U.S. Army's 2006 Counterinsurgency Doctrine' in *Oslo Files on Defence and Security*, 5, 2007, pp. 10–13.
25 Emiliy O. Goldman, 'The Spread of Western Military Models to Ottoman

Turkey and Maiji Japan' in Farrell and Terriff (eds), *The Sources of Military Change.*

26 David H. Ucko, *The New Counterinsurgency Era. Transforming the U.S. Military for Modern Wars*, Washington: Georgetown University Press, 2009. See also chapter 9 (Farrell) in this book; Stephen Biddle, 'Iraq, Afghanistan, and American Military Transformation' in John Baylis *et al.*, (eds), *Strategy in the Contemporary World*, Oxford/New York: Oxford University Press, 2nd edn., 2007, pp. 275, 291–2; David E. Johnson, *Learning Large Lessons. The Evolving Roles of Ground Power and Air Power in the Post-Cold War Era*, Santa Monica: RAND Corporation, 2007.

27 A new department had been established at the Staff College in the late 1990s for the sole purpose of doctrinal production, 'The Department of Military Force', and the 2000 doctrine was written by this department.

28 The Ministry of Defence was by that time organised more like the British MoD. In a Norwegian context this was a completely new way of organising the Central Organisation of Defence. It was called 'Integrated Strategic Leadership'.

29 *Strength and Relevance. Strategic Concept for the Norwegian Armed Forces*, Oslo: The Ministry of Defence, 2004. See also note 3 above.

30 St. prp. nr. 42 (2003–2004), *Den videre modernisering av Forsvaret i perioden 2005–2008* [White Paper No 42: *Further Modernization of the Armed Forces in the period 2005–2008*], paragraph 5.3.1. In Autumn that year the Ministry of Defence formulated/presented the task of doctrine revision in its letter of intent ('iverksettingsbrev') to the Chief of Defence, and by that the Chief of Defence had political backing for the doctrine revision.

31 Ibid. It must be noted that the MoD's reference to 'lessons learned' probably reflected an ambition to involve the Armed Forces' 'Centre for Military Experience' in doctrinal development. As we shall see this was not realised. The same can be said about the reference to 'CDE' (i.e. the Norwegian Battle Lab and Experimentation, NOBLE). On the contrary, 'defence technology' (i.e. the Defence Research Establishment) was involved to some extent.

32 MoD Archives 2004: Letter of 6 October 2004 (2004/069403-001/FST/31) from CHOD to the Defence Staff and the Norwegian Defence University College (DUC).

33 The quadrennial defence reviews, or 'long term plans' as they were called, were drawn up within a complex 'web' of political, military and other players. The Defence Research Establishment (i.e. the 'Norwegian RAND') contributed with so-called *Defence Analysis*, assessments based on quantitative methods. Moreover, from 1985 *The Chief of Defence's Long Term Defence Study* was published for the first time and subsequently repeated in 1991, 1996, 2000, 2003 and 2007. Together with the Defence Analysis from the Defence Research Establishment, the CHOD's Defence Study became the most important input in the Ministry of Defence's process of compiling the long-term plan. The final document was then presented to Parliament by the Ministry as a White Paper. After Parliament had discussed the White Paper and made the final changes it was sent back to the Ministry, which then had to implement the final version. This implementation was carried out by means of a letter of intent from the Ministry of Defence to the Chief of Defence.

34 *Forsvarets* [The Armed Forces' Joint Operational Doctrine, part A: Fundamentals], Oslo: Forsvarets overkommando, p. 3.

35 MoD Archives 2004: Letter 6 October 2004.

36 Olav Riste, *Norway's Foreign Relations*, Oslo: Universitetsforlaget, 2005, pp. 254–6. Norway wants to share its values and ideas with other countries based on the conviction that Norway can make a valuable contribution to international development, an aspiration which is perhaps not all that uncommon in small states.

37 MoD Archives 2004: Letter dated 6 October 2004.

38 *Joint Vision 2010. America's Military: Preparing for Tomorrow*, Washington D.C., Pentagon: Chairman of the JCS, 5126 Joint Staff, 1996. Updated and reissued May 2000, introducing the idea of 'full-spectrum dominance', see Barry R. Posen, 'Command of the Commons: The Military Foundation of U.S. Hegemony' in *International Security*, 28:1, Summer 2003, pp. 5–46.

39 Effects Based Approach to Operations (EBAO) is NATO's further development of the original American concept of Effects Based Operations (EBO). The United Kingdom has been the driving force behind the concept of EBAO. The European EBAO variant embodies a broader and more generally based approach than the original American concept with its strong emphasis on technology and air power. Hence, the concept of EBAO can be seen as a military interpretation of the comprehensive approach, closely linked to how the U.N. and the E.U. have been working to implement various versions of this approach.

40 MoD Archives 2004: Letter dated 6 October 2004.

41 *Allied Joint Publications* (AJP's) and *NATO Bi-SC: Guidelines for Operational Planning*, Brussels: NATO Doctrine and Publication Hierarchy; *JDP 01: Joint Operations*, London: U.K. Ministry of Defence, 2004; *The U.K. High Level Operational Concept*, London: U.K. Ministry of Defence, 2003; *British Defence Doctrine*, London: U.K. Ministry of Defence, 2001.

42 Bjerga and Gjeseth, ['The Home Guard and the Army: a Divided or Unified Territorial Defence?'], pp. 56–7.

43 MoD Archive 2004: Letter dated 6 October 2004.

44 Lieutenant Colonel Harald Håvoll, interview 9 December 2009.

45 The new Integrated Strategic Leadership of the Ministry of Defence (from 2003) implied that approximately half the high-level servants within the Ministry were officers.

46 Some of the original members of the Editorial Board left during the writing process because they were posted to other parts of the military organisation. They were replaced by other officers, but at no point did any naval officers join the board. However, two of the original members, both Air Force officers, including the Head of the Department of Military Force at the Command and Staff College, stayed on the Editorial Board until it was replaced by a smaller Writing Committee in Autumn 2005.

47 Lieutenant Colonel Harald Håvoll interview 9 December 2009.

48 Ibid.

49 One of the authors of this chapter, Kjell Inge Bjerga, was the civilian member of the new Writing Committee. The other member was Lieutenant Colonel Magnus Eriksson, who had become a member of the Editorial Board in the Spring of 2005 (before the disintegration). The Lieutenant Colonel had, of course, military education and experience, but had not participated in real operations. In 2006 the new Head of the Department of Military Force at the Command and Staff College, Lieutenant Colonel Professor John Andreas Olsen, also got involved in the writing. Olsen is an international expert in the field of American Air Power and Effects Based Operations. See: John Andreas Olsen, *Strategic Air Power in Desert Storm*, London: Frank Cass, 2003; John Andreas Olsen, *John Warden and the Renaissance of American Air Power*, Washington D.C.: Potomac Books, 2007.

50 Norwegian Defence University College (DUC), the Department of Military Force, Archive: *Catalogue with inputs from the 2006 remit*, dated November 2006.

51 On top of the document hierarchy making up the chain of command was the Chief of Defence's *Strategic Directive for Operations*, issued April 2003 (classified, see footnote 4), updated and reissued 15 December 2009.

52 Norwegian DUC Archive: *Letter from the Logistics Organisation (FLO),* dated October 2006. Quotation translated from Norwegian.
53 Norwegian DUC Archive: *Letter from the Intelligence Service,* September 2006. Quotation translated from Norwegian.
54 Norwegian DUC Archive: *Letter from the Medical Service,* dated October 2006. Quotation translated from Norwegian.
55 Norwegian DUC Archive: *Letter from the Air Force Academy,* dated 31 August 2006.
56 Norwegian DUC Archive: *Letter from the Air Force Staff,* dated 2 October 2006.
57 Ibid.
58 Norwegian DUC Archive: *Letter from the Army's Transformation and Doctrine Command,* dated September 2006.
59 This view reflected a deeply rooted cultural difference between the Army on the one hand, and the Navy and Air Force on the other. While the Army has been promoting joint operations and jointness between the single services in general since the World War Two, the Navy and Air Force have remained much more sceptical towards the idea of jointness as such. The main reason for this has been institutional interests: joint staffs have historically been dominated by the Army because of the fundamental asymmetry in numbers between the Army and the two others. See: Kjell Inge Bjerga, *Enhet som våpen* [Unity as a Weapon], Bergen: Eide, 2002, p. 14. For a broad discussion of diverging cultures within a military, see: Elisabeth Kier, *Imagining War. French and British Military Doctrines between the Wars,* Jersey: Princeton University Press, 1997, p. 19.
60 Norwegian DUC Archive: *Letter from the Naval War College,* dated September 2006.
61 Odd Arne Nissestad, *Leadership development: an empirical study of effectiveness of the leadership development program at The Royal Norwegian Naval Academy and its impact on preparing officers to execute leadership in today's conflicts and the conflicts in the years ahead,* Bergen: NHH, 2007.
62 Norwegian DUC Archive: *Letter from the Department of Joint Operations,* dated 29 September 2006.
63 Hew Strachan, 'Norwegian Armed Forces Joint Operational Doctrine' in *Norsk Militært Tidsskrift,* 4, 2009, p. 47.

8 The revolution in military affairs of the 'other side'

Itai Brun and Carmit Valensi[1]

It is customary to assume that since the 1970s there have been three major milestones in military innovation in the West. During the 1980s military theory and practice revolved around concepts such as ALB (Air–Land Battle) and FOFA (Follow-on Forces Attack); in the 1990s, discussions revolved around the RMA concept (Revolution in Military Affairs); and in the first decade of the twenty-first century the discourse on Transformation and its components, such as EBO (Effects Based Operations) and NCW (Network Centric Warfare), has attracted the most attention. The hub of these conceptual discussions was in the U.S., yet the implications were broader, and at the same time similar discussions were held in security establishments in other countries, including Israel.

This study argues that, in parallel to the developments in the West over the last three decades, a number of nations and organizations on 'the other side of the hill' have also undertaken a significant development in their military thought. One version of the outcomes of this development was recently described by Hassan Nasrallah, the leader of Hezbollah, in the following way: 'A new school of warfare, it doesn't resemble anything we know'.[2] This conceptual development is referred to in the study as the 'Other RMA' or 'O-RMA'.[3] This study aims to identify and describe O-RMA, to analyze the learning process that led to it and to trace its intellectual origins. The study concentrates mainly on the Middle East and, more specifically, deals with a group of states, including Syria, Iraq and Iran, as well as a group of non-state entities, including Hezbollah, Hamas and Al-Qaeda.

Since the late 1990s a comprehensive discussion relating to changes in the nature of the opponents and to the broader changes in the nature of warfare has been developed in the West. Nevertheless, it seems that a more systematic approach to the phenomenon referred to in this study as O-RMA is lacking. This study describes the unique 'learning community' that evolved around this innovation. It maintains that the current tactical patterns of these states and other entities share a broad common denominator that is not coincidental. These tactical patterns of warfare stem directly from common strategic and operational concepts that were developed during the 1990s and have been evolving since then. The type of warfare Israel encountered in Lebanon (in 2006) and in the Gaza Strip (in 2009) is a clear and challenging outcome of the conceptual development referred to in this study as O-RMA.

At the heart of this study stands the claim that the roots of O-RMA lie in a series of dramatic and tumultuous events that took place in the Middle East between the years 1979 and 1982. These events (the Islamic Revolution in Iran, the Soviet invasion of Afghanistan, the commencement of the Iran–Iraq war and the Israeli invasion of Lebanon) changed the political and social order in the region, and led to the formation of a 'radical axis' and a 'learning community' of states and organizations that are in conflict with the United States and Israel. Even though not all the elements in this radical axis share the same ideological and political vision, they do share the operational and strategic concepts that lie at the center of the conceptual development, and therefore they employ similar tactical patterns.

The study focuses on a time span of four decades, starting from 1979 until 2009. However, the latest upheaval in the Middle East, the so-called 'Arab Spring' (2010–2012), has been shaking the 'old order' that characterized the region. It seems that the current turmoil has a similar extent of impact as the one that occurred between 1979–1982, and may lead to another fundamental change in the strategic and operational environment.

This study makes use of primary materials in Arabic and Farsi that reflect the development of O-RMA, the threat assessment underlying it and the perspective of its developers on the process that led to it. The study is based on a wide selection of statements, speeches, books, fatwas, letters and other documents that have been released over the past three decades. The textual approach is complemented by a detailed analysis of the modes of operation and force design practiced by these elements. This enables an understanding of O-RMA, integrating theory and practice.

Upheaval and crisis: 1979–1991

Military historians and practitioners describe the influences of the Yom Kippur War, in October 1973, on the doctrinal developments in the U.S. Army during the 1970s, and especially on the doctrine of ALB. Yet it is clear nowadays that when General Don Starry and other officers learned the lessons from the Israeli battles against the Syrians in the Golan Heights, winds of change were already blowing in the Middle East. These winds would, in turn, lead some regional elements to develop a form of warfare that would turn out to be very different from the one carried out by Syria in 1973.

Upheaval: 1979–1982

Between 1979 and 1982, when the ALB doctrine was formed and the FOFA concepts were articulated,[4] four dramatic events shook the Middle East.

The Islamic revolution in Iran turned secular and pro-Western Iran into a radical Islamic Republic under the guidance of a Shi'ite cleric. The government's policy separated Iran from its weapon and doctrine suppliers in the West, severed relations with the United States and Israel, and partly isolated Iran. These moves proved very costly during Iran's war with Iraq. The establishment of the revolutionary regime also incorporated the notion of 'exporting the revolution' as an important element in Iranian strategy. The revolution inspired various Islamic movements, including Hezbollah and Hamas, and it later established Iran in a leading position in the Middle Eastern 'radical axis'.

The Soviet war in Afghanistan began with the Red Army's invasion in December 1979, and ended in February 1989 with the defeat of the Soviet Union. The Soviet involvement transformed internal conflict in the country, between the communist regime and the Islamic opposition, into a war of independence against a foreign power, and according to the views of Muslims in the area, necessitated a 'Jihad' for the liberation of Afghanistan from a heathen regime and a foreign occupier. Numerous volunteers from all over the Muslim world joined this Jihad, and military and financial assistance was provided to the Mujahedin camps. The defeat of the Soviet Union generated a 'victory narrative' for the Mujahedin, who fought the regime and the Soviet occupiers and created a generation of skilled and highly motivated warriors. Al-Qaeda and its associated movements are in many ways the outcome of this war.

The Israeli invasion of Lebanon in June 1982 led to the expulsion of the PLO from Lebanon and consequently to a fundamental change in its stand regarding the armed struggle against Israel. The war also revealed to the Syrians their military inferiority, and temporarily undermined Syrian hegemony in Lebanon. This state of affairs led to the vacuum that enabled Iran to take advantage of developments within the Shi'ite population in Lebanon, and to establish Hezbollah. The war led the Syrians to adopt, for a number of years, the 'strategic parity' doctrine, during which the Syrian army expanded significantly. Towards the mid-1980s Syria was in the midst of a grave economic crisis, which stemmed, to a large extent, from the priority given to military investments in an attempt to close the gap with Israel's technological superiority.

The Iran–Iraq War of 1980–1988 shaped Iranian strategy for the following two decades, greatly influencing its build-up of forces. The trauma caused by the war was the most significant factor in leading Iran to invest great resources in the development of its strategic force, with an emphasis on unconventional capabilities and ballistic missiles. Iran's failures in the war against Iraq taught Iranian leaders the importance of military organization and professionalism. The war greatly influenced Iraq's doctrine and force build-up as well. This was clearly manifested in the First Gulf War and the decade that followed.

Another meaningful event that took place during this period was the signing of the peace accord between Israel and Egypt, on March 26, 1979. The signing of the accord removed the Egyptians from the war cycle with

Israel, and left Syria on its own. This feeling of solitude had a major influence on the Syrians, whose strategic concepts always relied on Arab solidarity as a foundation for its security doctrine.

On June 7, 1981, the IAF attacked the nuclear reactor that was built in Baghdad (Osiraq). The attack revealed Israel's precision attack capabilities and affirmed Israel's readiness to use military force against threats to its national interests. The strategic Memorandum of Understanding, which was signed between Israel and the United States in the same year, had a great impact on the formation of the conception that the United States and Israel act in order to achieve similar interests, and that the two countries use similar weapons.

These events formed the strategic environment in which O-RMA evolved. In retrospect a more profound examination of the biographies of players such as Ayman Al-Zawahiri, Osama bin Laden,[5] Imad Mughniyah,[6] Abbas Musawi, Hassan Nasrallah,[7] Sheikh Ahmed Yassin and others reveals that the period between the end of the 1970s and the beginning of the 1980s formed, to a large extent, their personal biography, and the worldview of a new generation of leaders, commanders and warriors who had great influence over the conceptual development on the 'other side of the hill'.

The years 1979–1982 thus appear, in retrospect, as the years during which a fundamental upheaval took place, an upheaval that led to the development of O-RMA. When the FM 100–5 of August 1982 was published in the United States, all of these events were already under way, and Saddam Hussein had been ruling Iraq since 1979. The ALB doctrine that stood at the core of the 1982 manual was not related to any of these developments.[8] Instead it addressed the European arena and the Soviet threat, which was identified by Starry and others in the 1970s as the major threat to the free world in the context of the Cold War.[9]

Crisis: 1983–1991

The outcome of the dramatic events that took place between the years 1979 and 1982 were evident during the 1980s and became fully apparent towards the end of the 1980s and the beginning of the 1990s. Indeed, at the beginning of the 1990s, when the first RMA memorandums were circulated,[10] the initial foundations for O-RMA had already been laid. By this time the collapse of the Soviet Union was an established fact. The main militaries in the Middle East, most of which were equipped with Soviet weapons and Soviet doctrine, recognized the way in which the U.S. military was employed during the First Gulf War. This state of affairs led to a painful process which would force them to drastically change their strategic and operational doctrines.[11]

The collapse of the Soviet Union led to the collapse of Syria's national security strategy and ended its efforts to reach a strategic parity with Israel.[12] At the same time the Syrians began to comprehend the technological and conceptual developments in the West. Israel's technological superiority over Syria had already been demonstrated in 1982, when the IAF waged a massive standoff precision attack, supported by advanced Electronic Warfare (EW), against the Syrian SAMs in the Baqa'a valley.[13] At the end of the decade Syrians dealt with the threat of PGM with the help of their Soviet advisors. And yet it was the participation of Syrian officers in the coalition in the First Gulf War that actually enabled them to acknowledge the real dimensions of the threat.[14]

The Iraqis were also aware, generally speaking, of the nature of the threat, on the eve of the First Gulf War. Yet their force design and their thinking patterns, which developed over the years within the concept of industrial warfare, did not enable them to realize the full meaning of the developments in the West. Saddam's hope that the West's technological superiority would not count for anything in a ground war against his forces was smashed to pieces within 100 hours.[15] The lack of a decisive victory and the early withdrawal of the western forces made Saddam believe that he was able to take on the biggest power in the world and the coalition that joined it.[16] Nevertheless, Iraq was indeed defeated in the Gulf War. It lost a substantial part of its army, had to disarm its unconventional weapons and was subjected to severe sanctions, including No-Fly zones that were established in the north and the south of Iraq.

The end of the Iran–Iraq war spelled defeat for Iran; after the war, Iran could not realize its ambitious plans to purchase conventional weapons owing to financial difficulties and American pressure on supplier-countries. Even though Iran did not take an active part in the 1991 Gulf War, the outcome reinforced the lessons the Iranians had learned from the Iran–Iraq war. The superiority of Western armies over third-world armies was made perfectly clear to the Iranians.[17]

The period lasting from the mid-1980s to the beginning of the 1990s represents the strategic and military weakness of the Islamic states in the region (mainly Syria, Iraq and Iran) in the face of what appeared to be a new and problematic world order, which included an increasing military threat to these states from western forces. Against this background of state-level military weakness there were already three organizations in the foreground, organizations that were established in the 1980s and which would lead the war against the United States and Israel in the first decade of the twenty-first century: Hezbollah,[18] Hamas[19] and Al-Qaeda.[20]

Hezbollah was established in Lebanon in 1982, and a series of events had turned it into the most dominant organization in the struggle against Israel. Towards the end of 1983 the organization began a series of terrorist attacks, which led to the removal of American and French foreign forces from Lebanon in the Spring of 1984, and later on to the Israeli withdrawal into the Security Zone in the Summer of 1985.

The Hamas movement was established in Gaza by Sheikh Ahmed Yassin, shortly after the outbreak of the First Intifada (the Palestinian uprising) in

December 1987. In December 1992 Israel deported hundreds of Hamas activists to Lebanon. Upon the return of the deportees, who had developed good relations with Hezbollah personnel as well as with members of the Iranian Army of the Guardians of the Islamic Revolution, Hamas grew stronger and set up in opposition to the PLO.

Al-Qaeda was founded in 1988. For the organization, and for an entire generation of Islamic youth, the war in Afghanistan, which ended in 1989, was a formative experience. They considered the Mujahedin's victory over the Soviet empire a cultural and ideological victory and a sign of their ability to stand against complex military challenges. For many Muslims the collapse of the Soviet Union immediately after its defeat in Afghanistan was a direct outcome of this war.[21]

In the West this state of affairs, from the end of the 1980s to the early 1990s, served as a background for an optimistic period, with Francis Fukuyama's 'End of History' theory as its symbol.[22] Ideas of globalization and world peace originating in shared global economic interests also found their way to the Middle East, but were received with much less enthusiasm. The reality in the region pointed in another direction, one that necessitated the formation of a new strategy and a different kind of warfare.

Learning, fighting and evolving: 1991–1999

Military historians describe the fascinating dialogue (in the 1970s and the 1980s) between American doctrines and Soviet ones, as well as the impact of mutual learning on the conceptions in both superpowers.[23] O-RMA, the subject of this study, was not developed by such a systematic and thorough process. The American and Soviet conception had some impact on it, yet, in retrospect, it could be argued that this conception did not evolve as a result of studying the ALB doctrine, or of reading the RMA documents or the Soviet MTR concepts.

Methods of learning

The development of O-RMA was the result of a deep crisis and limited possibilities. It was influenced by the reality that created the American (and the Israeli) conceptions, yet it was formed on the basis of a parallel, but sometimes radically different, interpretation of these conceptions. O-RMA was propagated by political leaders and military officers in the Middle East, who acted according to their own cultural and historical heritage. Their point of view was also based upon close observation of the military operations that had taken place in the 1990s. More than anything else O-RMA was influenced by these political leaders and military officers' understanding of the social and political developments of the 1990s, a period during which their opponents, and especially the United States and Israel, had to conduct what David Halberstam has referred to as 'war in a time of peace'.[24] Thus the conception of the 'other side of the hill' was the product of three main learning techniques:[25]

First, learning the West's military doctrines. This was based mainly on observation of military operations in the 1990s as presented by the media (CNN). The study of military doctrines did not concern itself with in-depth details, but rather with understanding general developments and basic conceptions. Second, interpretation of the reality in the 1990s, from a religious, cultural and Islamic perspective: in this context changes in the global and regional order were emphasized, as well as social and economic changes, changes in military and civilian technology, and ideological changes. Third, friction with opponents on battlefields in Lebanon, in the Palestinian arena, in Iraq and other locations: this learning method was clearly evolutionary and led to the survival and development of warfare methods that proved to be more efficient.

Threat assessment

The First Gulf War provided an initial and satisfying picture as to the future direction of the forthcoming Revolution in Military Affairs. The war exemplified the major progress that the United States had made in military affairs, and enabled different players in the Middle East to update their intelligence assessments regarding the threat they would have to face.

Regarding Coalition air power, it was evident that its survivability and capability of penetration into enemy territory were greatly enhanced, due to developments in electronic warfare, stealth technology, unmanned air vehicles and standoff weapons. The striking capabilities of the Coalition air forces were improved significantly, due to guided missiles, better intelligence assets, and new command and control systems. These developments enabled the Coalition air forces to execute massive attacks on a large number of Iraqi targets, stationary as well as mobile, at any range, in any weather or light conditions, and in every arena. Regarding the Coalition ground forces, it appeared that their maneuvering capabilities into and within enemy territory had improved significantly.

In the absence of quality intelligence, and in light of the perceived link between the United States and Israel, Middle Eastern elements attributed to Israel American military capabilities, and formed their threat assessment accordingly.

Still, there were also some encouraging signs for those who were able to maintain an appropriate perspective. First and foremost, there was no clear-cut victory in the end. Saddam remained in power and Iraq maintained its territorial integrity. In many respects this was a clear example of what would be referred to years later by Western observers as 'victory by not losing'. In the more operational aspects, the war exemplified the importance of surface-to-surface missiles, and the West's lack of capability to cope with this challenge.

Conflicts: 1992–1999

The First Gulf War was one of the first military conflicts of the 1990s, and it served as a testing ground (both theoretically and operationally) for future conflicts. In the discourse concerning the development of O-RMA we can identify references to the following conflicts: the American operation in Mogadishu in 1993 ('Gothic Serpent'); the first war in Chechnya between 1994–1996; the punitive operations in Iraq following the First Gulf War and especially the 'Desert Strike' and 'Desert Fox' operations (1996 and 1998 respectively); the air strikes on Serbia in 1995 ('Deliberate Force'); the war in Kosovo ('Allied Force'); attacks in Sudan and Afghanistan in 1998, which took place as a reaction to the terrorist attacks on embassies in Kenya and Tanzania ('Infinite Reach'); the ongoing Israeli conflict in Lebanon with an emphasis on the military operations that took place in 1993 ('Operation Accountability') and 1996 ('Grapes of Wrath'), and the ongoing Israeli–Palestinian conflict, starting with the First Intifada at the end of 1987 and continuing with the terrorist attacks of the mid-1990s.[26]

Weaknesses of the RMA

One of the most conspicuous characteristics of conflicts in the 1990s was the almost absolute priority that was given to air power and the use of cruise missiles, while avoiding the use of ground forces. In the Gulf War there was a 39-day air strike prior to the ground operations; only airplanes and missiles were used during the punitive operations in Iraq, and in attacks in Sudan, Afghanistan and Kosovo; in Lebanon, Israel also primarily used its Air Force. These operations reflected the impressive improvements that had been made in air power abilities, yet they also highlighted the disadvantages that these abilities engendered. Air power indeed has many advantages, but the decision to rely mainly on these abilities also revealed its shortcomings – especially the increasing reliance on precision weapons, and the dependency on precise intelligence and suitable weather conditions.[27]

In Iraq, Kosovo and Lebanon, it turned out that Western technology was sensitive, first and foremost to target intelligence. The first signs of this already appeared in the utter failure to locate the surface-to-surface missiles in western Iraq in the First Gulf War. Israel had a similar difficulty with its attempts to locate Hezbollah's rocket launchers throughout the decade, and especially during the intense military operations in 1993 and 1996.[28] Four days of consecutive air strikes in Iraq during operation 'Desert Fox' made it clear that modern technology finds it hard to deal with an experienced opponent who has intelligence regarding the attack and acts to conceal his strategic assets. A similar situation was also apparent in the other punitive operations in Iraq during that decade, as well as the attacks by cruise missiles in Sudan and Afghanistan. Military forces that retreated from Kosovo following the Allied Force operation exemplified these difficulties and showed that, once an opponent was prepared for an air strike, it could minimize its effects.

The conflicts clearly reflected the weaknesses associated with Western leadership: that it is highly sensitive to public opinion and is influenced by a number of elements, notably military casualties, civilian casualties (on both sides), as well as the overall cost of military operations. The military operations of the 1990s showed clearly that the West preferred a quick war that ended in conclusive victory with minimum casualties. All of the military operations in the 1990s were carried out according to a strict policy that minimized the risk to the attacking forces. Such a policy led to the use of air forces during the hours of darkness; avoiding attacking in areas of high risk; giving priority to precision weapons launched from safe areas, and, especially, flying high above the threat areas.[29]

To a deeper extent the military operations reflected an increasing gap between the traditional image of decisive military victory and the actual ability to achieve it using the military doctrines that were formulated throughout the decade. Opponents realized that the West was held captive by its own image regarding the manner in which military operations ended, an image that was mostly created by the image of the ceremonies of victory and surrender in World Wars I and II and the wars between Israel and the Arabs. The gap between the actual outcome of the wars and operations and the image that accompanied them had tremendous impact on the development of O-RMA.

The O-RMA

Thus a military doctrine was shaped, with the underlying assumption that while one side is technologically superior, there could be parity and equality between the two sides in other areas, and the scales might also be turned to favor the technologically weaker side. The latter could, for example, have more territories, higher populations, and even enjoy a number of tactical and strategic advantages. Moreover, the asymmetry between the sides was not only limited to material issues. The underlying assumption of this military doctrine is that there could also be asymmetry in the importance of the interests each side has, the objectives of the war, the level of determination and resolution, willingness to take risks, and sensitivity to casualties. The technologically inferior side is also free from political and cultural constraints that limit the stronger side.[30]

Towards the end of the 1990s most of the central players in the Middle East were already undergoing an intensive process in which they followed three main paths as a result of the new military reality.

The first path was the need to significantly improve their ability to withstand attacks, in order to prolong absorptive capacity and preserve strength. This was borne out of an awareness of the lethality of precision weapons and their influence upon the battlefield. From the start the underlying assumption regarding the improvement of absorptive capacity was the Soviet doctrine which held that appropriate preparedness for absorption would lead the West to waste its precision weapons, leading to a situation in which the latter stages of the conflict would be fought according to older patterns. Nevertheless, as time progressed, it appeared that sustainability and preservation of power had other advantages as well.[31]

The second was the need to establish a credible deterrent capability, first and foremost to prevent a comprehensive conflict that would be contrary to their interests and their capabilities, but also to lead the stronger side to areas more favorable for the weaker side should basic deterrent strategies fail. Such a move curtails some of the technological advantages of the attacker.[32]

The third and final path was the need to move from a decisive win strategy to a strategy of attrition. This is considered by all the elements as key to victory, and is a result of Western sensitivity to continuous war and casualties.[33] Attrition is meant to lead the opponent to believe that the situation is at a dead end, and, consequently, it will give up and leave. This led to the concept of 'victory by not losing'.[34] According to this concept surviving the conflict is a key factor in winning it, and it is based on the West's inability to achieve a clear-cut and decisive victory.

At the heart of O-RMA there is an acknowledgment of the technological supremacy of the opposing side. But at the same time it is based on the assumption that the stronger side also has weaknesses that can be targeted in ways that will negate its technological supremacy. Based on these assumptions a type of warfare developed with the following six components:

First, a strong emphasis on the survivability of the fighting forces and other systems, as part of the overall absorptive effort and sustainability: the improvement of survivability is achieved by the use of protective means (bunkers and especially tunnels), camouflage and deception, scattering military forces, deliberate obfuscation of military and civilian facilities, and conducting the war in an urban area filled with civilians and the media. There is also an extensive use of low-signature systems (personal anti-tank and anti-aircraft guided missiles, and surface-to-surface rockets) and low signature forces (commando units, infantry, guerrilla warriors, paramilitary forces, suicide bombers), and the adoption of warfare methods that preserve such a signature (especially terror and guerrilla warfare).[35]

Second, a strong emphasis on high-trajectory ballistic weapons (SSMs and SSRs), whose advantage is their technical simplicity, low cost, and ability to penetrate deep into enemy territory without encountering an adequate response: it is difficult to locate such launchers, owing to their low signature and vast numbers. Within this frame of reference a variety of rockets and missiles were developed and purchased, ranging from simple and improvised rockets with a range of a few kilometres, to long-range missiles such as the Iraqi 'Al Hussein' missile (650 km) and the Iranian 'Shihab 3' (1,300 km). Rockets and missiles are operated from stationary and mobile launchers that are sometimes camouflaged in civilian facilities and vehicles. Rockets and missiles are important in terms of both deterrence and attrition efforts.[36]

Third, a strong emphasis on the use of weapons and methods of operation that can lead to a high number of casualties among civilians and army forces: within this framework suicide bombers and a variety of explosive devices are used, as well as IEDs. The choice of suicide bombers as a central component of asymmetrical warfare is a result of the high availability of volunteers, and the impact this method has on morale and on the media. The idea of inflicting casualties is a basic and longstanding one in warfare. Still, in asymmetrical warfare there is no intention of eliminating the opponent's forces, but of making use of his sensitivity to casualties in a manner that will create pressure to end the war, accompanied by a feeling of failure. The number of casualties is important in the context of deterrence as well as attrition.[37]

Fourth, a strong emphasis on media and propaganda efforts vis-à-vis the local population, the opponents' population and the international community: this kind of effort aims to challenge the legitimacy (internal and international) of the West's military operations. This is achieved by demonstrating the heavy price that comes with this. The media efforts shape an internal and international agenda and undermine public resilience in the opponent's country.[38]

Fifth, a strong emphasis on the effort to force the war into 'close battle': the underlying assumption of all the players is that in such battles the technological supremacy of the opponents will be significantly undermined.[39]

Sixth, a strong emphasis on confronting the opponent's aerial supremacy with the use of active means (aerial defence and attack systems) and passive ones (as part of the entire absorptive effort): this emphasis had a significant influence on the manner in which forces were deployed, and the way they operated. In this context this issue has led some of the other players to invest in their power structure, purchase new aerial defense systems and upgrade existing systems.[40]

A different direction, one that exceeds the scope of this study but is closely related to it, is the non-conventional direction, with an emphasis on nuclear technology. Iraq was a pioneer in this area even as early as the end of the 1970s, and it was followed by Iran, Libya and probably Syria.

Roots

O-RMA is an eclectic concept whose origins can be traced to a variety of sources: religious and secular, Eastern and Western, theoretical as well as practical.

War in Islam and the legacy of Jihad

Warfare occupies a central position in Islam, in which the struggle between good and bad was traditionally translated into military and political dimensions.[41] Muhammad himself was not only a prophet and a teacher, like the founders of other religions, he was also a ruler and a warrior. The rhetoric that accompanies O-RMA has a central place held by concepts, quotes, events and ideas that originate in the Qur'an and in Islamic history and tradition. The image that became so prevalent in the final decade, of the suicide bomber holding the Qur'an, is a visual expression of this central phenomenon and reflects the three roles played by Islam in the development of O-RMA.

First, there is the use of religious texts in order to recruit fighters and to generate wide public sympathy concerning the entities and states involved.[42] Second, there is the use of religious texts in order to communicate with fighters, as well as instructions regarding tactical patterns of warfare and the ideas underlying them.[43] Third, texts are a fundamental source of inspiration to the conception of warfare itself.[44]

The concept around which the development of O-RMA mainly revolved was 'Jihad'. This concept received a new interpretation during the twentieth century. The interpretation was developed by a series of thinkers who belonged, during different periods of time, to the Muslim Brotherhood movement. Among these thinkers we should mention Hassan al-Banna (1906–1949), the founder of the Muslim Brotherhood movement, Sayyid Qutb (1906–1966), al-Banna's successor in the Muslim Brotherhood movement, and Abdullah Yusuf Azzam (1941–1989), one of the founders of Al-Qaeda.[45]

The PLO legacy

Many of the roots of O-RMA, especially in its strategic aspects, are to be found in Palestinian armed resistance as it developed during the late 1950s and the early 1960s. This conception was revolutionary at the time. Contrary to the basic assumption of Nasserist strategy, according to which, as long as Arabs are inferior in their powers to Israel they should not act in a manner that might provoke Israel to retaliate, the founders of the Fatah claimed that Israel's technological supremacy would be undermined in a long process of attrition that was based upon guerrilla warfare. They also claimed that guerrilla operations would not provide Israel with a reason to wage an overall war.

The relations that were established by Iranian elements that opposed the Shah and the PLO, alongside the transfer to Hezbollah of Lebanese fighters who were members of the PLO, led to Hezbollah's adoption of Palestinian warfare methods. Imad Mughniyah, who was a member of the PLO's elite unit ('Force 17'), played a central role in this regard.

The Guerilla Legacy (and the 'Muqawama')

In a more operational context, many of the relevant entities characterize their warfare method as guerrilla warfare. Some studied guerrilla warfare independently, but in this context it seems that some of the origins of guerrilla warfare are related to its heyday in the 1950s and the beginning of the 1960s in China, Cuba, Algeria and particularly Vietnam. The Palestinians translated guerilla literature into Arabic and were influenced by the ideas expressed in it. History proved that guerilla warfare was a method that could bridge the gaps between a superior opponent and a local, low-budget independence movement.[46]

The concept that connects the guerrilla legacy with the more general ideas that stand at the centre of O-RMA is 'Muqawama' ('El-Muqawama') – Resistance. Over the past few decades the concept gained unprecedented influence in the Middle East. Within this frame of reference there was an almost total overlap between the concept of resistance and fundamentalist Islam, and resistance gained in popularity along with fundamentalism.

Foreign influences

The influence of more institutionalized military thinking on the 'other side of the hill' can also be detected. The Soviet Union was the main munitions and military knowledge provider to Middle Eastern countries, and especially to Syria and Egypt (until the Yom Kippur War).[47] And yet, as O-RMA began to evolve, the Soviet Union was already heading towards collapse.

Therefore Soviet military doctrine, and later on the Russian doctrine, only influenced the Syrians for a relatively brief period at the end of the 1980s and the beginning of the 1990s. Even prior to the Gulf War Soviet counsellors taught the Syrians about the fundamental change that had taken place on the battlefield with the entrance of precision weapons.

The Soviet response to this change, which included both an active component (targeting precision weapons) and a passive one (camouflage, deception, defense and the dispersion of military forces) had a great influence over the Syrians and other regional players.

The export of weapons from North Korea to the Middle East commenced in the 1980s; it included the provision of weapons to various countries, especially Iran and Syria. North Korea had a great influence on the missile and rocket systems of these countries. It provided Iran with Scud missiles during its war with Iraq, and after that it played a central part in the establishment of Iran's missile infrastructure, as well as developing its missile systems. Similar aid was also given to the Syrians and there are also reports concerning shipments of weapons to Hezbollah. It is likely that, along with such shipments, the relevant knowledge was also transferred regarding the operation of such weapons and the manner by which they can be assimilated in the overall warfare conception.

Lessons from the Yom Kippur War

Another significant inspiration for O-RMA concerns the more institutionalized lessons that were learned by Arab countries from the 1973 Yom Kippur War. The Arabs were less impressed by the heroic tales of Generals Tal, Eitan and Peled, that stirred the minds of Don Starry and other Americans.[48] Despite three humiliating defeats the Arabs actually considered the Yom Kippur War a clear victory. Two conflicting lessons were learned from the war: first, a war can be waged on Israel that results in significant gains; and second, despite ideal conditions at the beginning of the war, the IDF could not be defeated.

The war strengthened the Arab understanding that Israel's weaknesses could be exploited in three main areas: causing as many casualties as possible (due to Israel's sensitivity to the value of human lives); undermining its home front and by so doing influencing the operational battlefield; and prolonging the war as long as possible, while maintaining the initial military achievements in a manner that would lead the IDF to make use of its means of defense rather than its means of attack.[49]

Lessons from the First Lebanon War

The lessons learned from this war strengthened the conception that to counter Israel's maneuvering abilities a large number and complex set of obstacles should be put in place, from the border up to the outskirts of Damascus. Such obstacles would include outposts, anti-tank ditches, landmines, ramparts, etc., with the support of huge quantities of anti-tank weapons. The results of the First Lebanon War showed the Syrians that their ground forces were well structured and well operated and that appropriate use of anti-tank weapons against tank brigades could have strategic implications.[50]

Lessons from Afghanistan and the Iran–Iraq War

But the wars that had the greatest impact on O-RMA were the one in Afghanistan and the Iran–Iraq War. During the war in Afghanistan Mujahedeen forces applied guerrilla methods, setting up ambushes and waging surprise attacks, especially on main transportation routes and secluded outposts. During their stay in Afghanistan, Mujahedeen volunteers underwent military training and acquired vast experience in guerrilla warfare.[51]

The war with Iraq was the source of many of the components that comprise O-RMA. These forms of warfare were learned and practiced by Iran and Iraq during the 1990s. The Iran–Iraq war served as the main catalyst to the development of a series of short and mid-range rockets. During the 1990s Iranian military industries developed the 'Fajr 3' and the 'Fajr 5' rockets that were delivered to Hezbollah. The war with Iraq was also the reason for purchasing ballistic missiles, as well as the huge investment that was made in their subsequent development.[52]

Reality check: 2000–2009

During the 1990s the elements with which this study deals did not achieve significant accomplishments. This state of affairs began to change in 2000 and 2001, when a number of significant events took place, events that were perceived by these elements as the realization or implementation of O-RMA. These events led to a series of statements in which O-RMA was presented as an efficient response to the technological supremacy of the United States and Israel.

Achievements: 2000–2001

The IDF's withdrawal from Lebanon in May 2000 was considered to be a great success for Hezbollah. The organization was perceived as having achieved the impossible – the humiliating expulsion of the IDF from southern Lebanon without any simultaneous concessions to Israel. Following the IDF's retreat the Secretary General of Hezbollah, Hassan Nasrallah, delivered a series of speeches in which he presented the ideas which underpin Hezbollah's warfare strategy, as a new and efficient strategy in the struggle against Israel and as a historic turning point in the struggle.[53]

The terrorist attacks on the Twin Towers in New York and on the Pentagon on September 11, 2001, were also considered a successful manifestation of the warfare patterns that had evolved during the 1990s. The fact that a few individuals, equipped with primitive means, turned passenger airplanes into guided bombs and hit important locations in the United States, excited many in the Muslim world. The attacks left the United States shocked, bewildered and in a state of collective pain.

The outbreak of the Second Intifada, in September 2000, and the activities that took place during its first years were also perceived as a success. A year after the outbreak of the al-Aqsa Intifada, the Hamas journal published a series of articles that analyzed the strategic significance of suicide bombings. These articles stressed the fact that suicide bombings not only created a 'balance of terror' with Israel, but also provided advantages to the Palestinians. Israel is described in these articles as lacking any deterrence or offensive capabilities. The articles also claimed that Israeli technological supremacy could not come to terms with the Palestinians' determination and willingness to sacrifice their men and women. A year later Khaled Mashal defined the strategic objectives of Hamas in the conflict as 'attrition' that was meant to 'make the Zionists doubt their own future'.[54]

Iraq 2003

After failing to prevent the war, Saddam Hussein anticipated that the key to his survivability was to conduct a long war with high casualties. This would arouse debate in the United States and lead other states to call for the war to end.[55]

Saddam acknowledged the absolute technological superiority of the United States, but his basic assumption was that an air campaign alone would not be sufficient for the Coalition to meet its strategic objectives. Therefore, for Saddam, a key aspect of the conflict was the need to preserve Iraqi forces' survivability during the air strike phase.

Another underlying assumption behind Iraq's strategy concerned its desire to channel the ground operations to the outskirts of its cities, and to conduct the war while assimilating its forces into civilian society and civilian facilities. Saddam's irregular forces were meant to conduct guerrilla warfare and create pockets of resistance, even in areas that were conquered by American soldiers.

The third component in the strategy was the need to conduct a war on public opinion in the media, with three target audiences in mind: the Iraqis themselves, the American population, and the international community with an emphasis on Europe and the citizens of other Arab countries. Saddam hoped that the image of a victim nation facing a savage attack against its civilians and its basic infrastructure would lead to international intervention and the end of the war before its main strategic objective could be achieved.

Contrary to Saddam's expectations the Americans made use of a strategy that did not necessitate a preliminary air strike. The American ground forces advanced more than 500 km in less than 20 days, with the loss of only 60 soldiers. Their arrival in Baghdad resulted in Saddam's defeat and demonstrated the utter failure of his warfare strategy. Nevertheless, the war did not end with the conquest of Baghdad, and soon the American forces in Iraq had to deal with a much more sophisticated version of O-RMA, which was carried out by a large number of groups and organizations.

The Second Lebanon War: 2006

Syria's and Iran's support of Hezbollah led to a situation in which Hezbollah, on the eve of the Second Lebanon War, was a unique example of a terrorist organization with significant military capabilities. At the beginning of the war it had a large reservoir of long-range rockets (up to 250 km); a very large quantity (13,000) of short-range rockets; an air force (attack UAVs), as well as naval capabilities (anti-ship missiles), and large ground forces (about 10,000 fighters) that operated like guerrilla forces and were armed with anti-aircraft and anti-tank missiles. Hezbollah had erected special facilities in southern Lebanon, in which it placed anti-tank weapons; underground bunkers were built, and a logistical infrastructure was created with the aim of carrying out lengthy attacks. Some of Hezbollah's facilities and buildings were on the outskirts of Shi'ite villages or in the villages themselves.[56]

The assumption underlying Hezbollah's force structure and its preparations for a possible conflict with Israel was 'victory by not losing'. This is a variation of the attrition concept, based on a deep understanding of Israel's core conceptions regarding its military and security. Israel needed a clear-cut victory in a quick war, and all that Hezbollah needed was to survive and demonstrate that survivability. This survivability was demonstrated by an ongoing barrage of rockets into Israeli territory, which was intended to cause the highest number of civilian casualties. At the same time this impeded the progress of the Israeli forces, increasing Hezbollah's ability to continue to fire rockets.

Israel has enjoyed a few quiet years on its northern border, a fact that is undoubtedly connected to the positive strategic outcome of the second Lebanon war. Nevertheless, general opinion in Israel considered this war to have been a missed opportunity. This is due to the feeling regarding the expectations gap that was created at the beginning of the war and the manner in which it ended. The IDF, in spite of its highly advanced capabilities, could not stop the rockets that were fired into northern Israel, and Hezbollah was not defeated. The 4,100 rockets that were fired into Israeli territory demonstrated the limits of military might, as well as the complexity of the O-RMA challenge.[57]

Gaza 2009

After Hamas took over the Gaza strip in June 2007, Iran and Hezbollah passed on the operational knowledge they had gained as a result of the war in Lebanon. Hezbollah's method of operation was a role model for Hamas. Iranian aid to Hamas between 2007 and 2009 consisted of the provision of various weapons that were smuggled into the Gaza Strip (especially rockets with a range of 20–40 km and anti-tank missiles) and technological skills that enabled the installation of explosive devices similar to the ones Hezbollah had used in the Second Lebanon War. Hundreds of Hamas members left the Gaza Strip and underwent advanced training in Iran, Syria and Lebanon. In addition, thousands of new recruits underwent training in the Gaza Strip itself.

The defensive conception regarding the Gaza Strip, developed by Hamas using aid from Iran and Hezbollah, was based on the idea of inflicting large numbers of casualties upon the IDF by avoiding open area conflict and conducting urban warfare in densely populated areas. Within such areas the IDF would encounter landmines, snipers, anti-tank missiles, explosive devices and suicide bombers. Hamas made its disappearance techniques more sophisticated, making extensive use of tunnels and assimilating its fighters into the civilian population. At the same time Hamas made preparations to wear out Israel's home front by launching rockets even while the IDF continued its military operations; in addition, Hamas made extensive use of the media in order to undermine the legitimacy of Israel's operation. Drawing on outcomes learned from the Second Lebanon War, Hamas estimated that continuous firing would undermine Israel's sense of military achievement, and lead to frustration and the feeling that the IDF had failed in its mission.

It is clear that Hamas's achievements were nowhere near those of Hezbollah in Lebanon in 2006. The number of Israeli casualties was very low (ten soldiers and three civilians) and the damage inflicted on Israel was very limited. The IDF made a ground forces maneuver into the Gaza Strip and several hundred Hamas members were killed during the operation.

Conclusion

The acknowledgment of the 'other side of the hill' of their technological inferiority and the large military gap between these states and entities and the West lies at the centre of the O-RMA. In light of this gap, these elements have established a method of warfare that is based on the assumption that alongside their superiority, the 'stronger side' also has weak points and vulnerabilities that can be exploited to offset technological superiority. This 'way of war' is based on the following components: improving absorption capability, in order to increase survivability and provide a breathing space for the 'weaker side', creating effective deterrence, in order to deter the 'stronger side' from attacking the 'weaker side' and shifting the war to more convenient areas in case this deterrent fails; and winning the war by not losing it, while creating an attrition effect.

The implementation of this concept leads to tactical patterns that stress the survivability of the fighting force and other systems, the use of ballistic weapons, and the use of weapons and methods that can lead to a large number of casualties among civilians and armed forces. There is the desire to turn the war into a close battle, with the need to cope, actively and passively, with the stronger side's aerial superiority. All of these elements have learned to make use of the media in order to increase the influence of their actions.

These various elements learned their lessons from the military conflicts in the 1990s and implemented these lessons in the creation of their threat assessment. The ongoing friction between some of these elements and the United States and Israel contributed considerably to the evolution of both the strategic and operational concepts and the tactical patterns.

O-RMA is an exceptionally eclectic conception, whose sources include, (among others) the classic guerrilla legacy, the influence of foreign militaries such as the Soviet Union and North Korea, lessons learned from the Arab–Israeli wars, the Iran–Iraq War, and the war in Afghanistan.

The complex outcomes of the conflicts in Lebanon (2006) and the Gaza Strip (2009), along with the current changes in the geo-strategic environment, require paying special attention to the question of how the different elements analyze the effectiveness of O-RMA in light of these conflicts. Both organizations, Hamas and Hezbollah, did survive the clashes with Israel and even though their military capabilities were harmed during the clashes they managed to restore it. Nevertheless, from a strategic point of view, both organizations were hit badly, which led to a fundamental change in their policy. More than five years after the war Hezbollah is avoiding waging terror attacks on Israel, a policy that is still adhered to at the time these lines are written. Hamas came out as a clear loser and it appears that it has adopted a survival policy that avoids waging any attacks on Israel in the near future.

The outcomes of those conflicts can lead to at least three different possible future trends that require careful study of those who are dealing with conceptual developments in the West. One possible trend ('more of the same') might be the continuation of O-RMA in its current guise. If so, the basic components of O-RMA (i.e. the strategic ideas of attrition, deterrence and absorption) will continue to serve as the basis for military thought among the group of actors studied here. A close but slightly different trend ('increasing lethality') might be the continuation of O-RMA's three major components, but a change in the tactical patterns that will develop in more lethal directions. A third trend ('a paradigm shift') might be a radical change of O-RMA in directions that can be contradicting to it. Possible change might be connected with the element of attrition. Another possibility that reflects a different direction for O-RMA is using the 'weapon' of diplomacy and propaganda to a greater extent ('a continuation of resistance by other means'). This effort aims to delegitimize states, as in the case of Israel. The purpose of this effort is to isolate Israel and to denounce its ability to defend itself properly. The 'Gaza Freedom Flotilla' of May 2010 is a clear example of this. The effort involves 'peace organizations' and 'human rights' organizations and has achieved some success, though it seems that following the first flotilla Israel has learned lessons and improved its public relations response.

So far the concept of O-RMA still serves as the basis for the force build-up for all the entities including Hezbollah and Hamas. However, those two organizations have changed dramatically since their establishment.

The current unique features of these organizations pose a question mark over the traditional definition of 'violent non-state actors'. Their uniqueness stems from their ability to hold several identities simultaneously. Identities hitherto associated either with state actors or non-state actors. In these cases, the actors combine identities of both worlds. In fact, they constitute a new type of 'hybrid actors' which blend four main identities: sub-state identity as in ethnic–community affiliation (Sunni, Shi'ite); supra-state identity of pan-ideology (pan-Islamic); state-like identity including implying state functions (government, infrastructures, military) and non-state identity as in organizational behavior (terror/guerrilla). These identities eventually led the organizations to simultaneously operate in various spheres: military, social and political.[58] The military implication of this 'hybridism' is the organizations' ability to employ a fused mix of conventional weapons, irregular tactics, terrorism and criminal behavior in the battle space, in order to obtain their political objectives.[59] However this 'hybridism' does not always work in the organization's favor and may generate tension between competing identities.

In this regard both Hamas and Hezbollah operate in the political sphere along with the military and the social one. This reality obliges them to maneuver between the different obligations. The political sphere became more important for them in the last decade, as they became governmental entities. The implication of this understanding is that they both have to demonstrate a greater accountability toward their constituency. They no longer have the same freedom to use their military forces without reference to the possible consequences of doing this. The trend of 'paradigm shift' may be enhanced in light of this understanding, and may lead to a possible shift from a strategy of attrition into a massive strike one. This will make it possible to shorten the operation and by that to minimize the losses and the collateral damage.

This chapter has been based on the assumptions that developments in the strategic environment have dramatic impact on operational concepts. In that sense the current wave of protests and demonstrations, called the 'Arab Spring', might bring about changes in the concept of O-RMA.

There is an important and relevant debate among researchers concerning the scope of the phenomenon. Although it is still early to tell how, it seems that recent events have in many ways changed the existing regional order. However, a 'new order' hasn't been formulated yet, hence, for the moment, the region goes through a transition period of instability and uncertainty.

The popular protests strongly stressed the importance of public opinion among the group of states and organizations. It seems that this factor will increase the sense of accountability toward the population and as a result will lead the entities to strive for shorter wars as a change from the attrition concept of O-RMA.

So, what next? This chapter has concentrated on the past, and with this historical approach it can tell us very little about the future. The only lesson that can be taken out of this chapter is the need to pay attention to the 'other side of the hill', and to consider its possible developments. It seems that the recent developments in the Middle East will not only change the strategic environment but also affect the operational one in a manner that needs to be studied. Listening to what the other side is saying (in private and in public) might help. Otherwise, Nasrallah's promise of 'great surprises that could change the fate of the war'[60] might be fulfilled.

Notes

1 Brigadier General Itai Brun and Carmit Valensi would like to thank Dmitry (Dima) Adamsky, Zehavit Berman, Haim Meyer, Ori Goldberg, Avi Altman and Yael Brahms for their assistance.
2 Hassan Nasrallah, speech, Al-Manar TV station, February 22, 2008.
3 We prefer the term, 'the Other RMA', as a general title for military innovations, without committing to the term 'revolution'. It seems that the term 'evolution' fits the way this chapter describes the process of O-RMA development more accurately.
4 Richard Lock-Pullan, 'How to rethink war: conceptual innovation and AirLand Battle Doctrine' in *Journal of Strategic Studies*, 28:4, pp. 679–702, August 2005.
5 On Osama Bin Laden and Ayman Al-Zawahiri, see: Lawrence Wright, *The Looming Tower: Al Qaeda and the Road to 9/11*, Knopf, 2006; On Al-Zawahiri, in his own words, see also: Ayman Al-Zawahiri, *Knights under the Prophet's Banner*, Dar-al-Najaah Al-Jadeedah, 2001.
6 On Imad Mughniyah, see: Ibrahim al-Amin, 'Things that can be said about the resistance magician' in *Al-Akhbar* [newspaper], February 12, 2009.
7 On Hassan Nasrallah and Abbas Musawi, see: Shimon Shapira, *Hezbollah between Iran and Lebanon*, [in Hebrew], Hakibbutz Hameuchad, 2000.
8 U.S. Department of the Army, *Field Manual 100–5: Operations*, Washington D.C., 20 August, 1982.
9 General Donn A. Starry, 'A Tactical Evolution – FM 100–5' in *Military Review* 58, August, 1978, pp. 2–11.
10 Andrew Marshall, *Some Thoughts on Military Revolutions – Second Version*, Washington D.C.: Office of the Secretary of Defense, 23 August, 1993.
11 For Arab military strategy and operational doctrine prior to the 1990s, see: Yehoshafat Harkabi, *Arab strategies and Israel's response*, New York: The Free Press, 1977.
12 On the impact of the collapse of the Soviet Union on the Syrian national security strategy, see: Eyal Zisser, *Assad's Syria at a Crossroads* [in Hebrew], 1999, p. 65.
13 Hafez al-Assad, *A Message from the President of the Republic and the Supreme Commander to his Fighting Sons*, June 21, 1982, cited in: Mustafa Tlas, *The Israeli Invasion to Lebanon* [Hebrew translation], 1988, p. 19.
14 See: Zisser, *Assad's Syria.*
15 Saddam Hussein, speech to the Islamic Conference in Baghdad, January 11, 1991.
16 Kevin M. Woods, *The Iraqi perspective report – Saddam's Senior Leadership on Operation Iraqi Freedom from the official U.S. Joint Forces Command report*, Maryland: Naval Institute Press, 2006, p. 16.
17 Ephraim Kam, *From Terror to Nuclear Bombs: The Significance of the Iranian Threat*, [in Hebrew], 2006, chapters 1, 3.
18 On Hezbollah's background, see: Naim Qassem, *Hizbullah – the story from within*, Saqi Books, 2005. See also: Augustus Richard Norton, *Hezbollah: A Short History*, Princeton: Princeton University Press, 2007.

19 On Hamas's background, see: Matti Steinberg, *Facing their Fate*, [in Hebrew], 2008; Shaul Mishal and Avraham Sela, *The Palestinian Hamas*, Columbia: University Press, 2000.

20 On the linkage between winning the Afghanistan war and the collapse of the Soviet Union, see, for example: Al-Zawahiri, July 9, 2005.

21 On Al-Qaeda's background, see: Lawrence Wright, *The Looming Tower: Al Qaeda and the Road to 9/11*, Knopf, 2006.

22 Francis Fukuyama, *The End of History and the Last Man*, New York: Free Press, 1992.

23 Dima Adamsky, *The Culture of Military Innovation: Comparing the Revolution in Military Affairs in Russia, the U.S. and Israel*, Stanford CA: Stanford UP, 2010.

24 David Halberstam, *War in a Time of Peace: Bush, Clinton, and the Generals*, Scribner, 2001.

25 For learning techniques, see for example: Abu Ubeida [spokesman for Hamas], *Hamas Website*, January 10, 2009; Osama Bin Laden [and others], *Jihad against Jews and Crusaders*, 1998; Gholam Reza [head of the Iranian Passive Defense Authority], interview, *Sobhe-Sadeq*, May 19, 2008.

26 Osama Bin Laden, *Declaration of War against the Americans Occupying the Lands of the Two Holy Places*, London: *Al-Quds al-Arabi* [newspaper], 1996; Naim Qassem, *Hezbollah – the story from within*, Saqi Books, 2005, pp. 109–12, 114–18, and also many of Nasrallah's speeches.

27 The weak points of air power are clearly specified in statements that were made by Saddam, Nasrallah, Bashar al-Assad and others. See for example: Saddam Hussein, speech to the Islamic Conference in Baghdad, January 11, 1991.

28 Hassan Nasrallah, interview, *Alhawadat*, March 19, 1999.

29 For the Iraqi perspective concerning the sensitivity of the West to casualties, see: Kevin M. Woods, *The Iraqi perspective report – Saddam's Senior Leadership on Operation Iraqi Freedom from the official U.S. Joint Forces Command report*, Maryland: Naval Institute Press, 2006.

30 One of the clear conclusions that can be derived from analyzing O-RMA rhetoric is the similarity between the way the different elements describe the technological gap and the ways they devise for offsetting their inferiority. See for example: Bashar al-Assad, interview, *Al-Sharak al-Awsat*, June 12, 2000; Naim Qassem, *Hizbullah*, p. 73; Ismail Haniyeh, *Hamas website*, January 5, 2009; Osama Bin Laden, *Declaration of War*; Saddam Hussein to Yasser Arafat, April 1990, quoted in: Woods, *The Iraqi perspective report*, p. 6; Mohammad Ali Aziz Jafari [IRGC commander], interview, *Mehr News*, September 3, 2007.

31 On absorption as a key element of O-RMA, see for example: Bashar al-Assad, interview, *As-Safir* [newspaper], February 22, 2002.

32 On deterrence, see, for example: Rahim Safavi [Khamenei's advisor for Iran's armed forces], interview, *Mehr News*, September 23, 2007.

33 On attrition, see, for example: Muhamad Def [and other senior members of Hamas's military wing], interview, Al-Jazeera T.V., July 4, 2006.

34 On 'victory by not losing', see, for example: Hassan Nasrallah, interview, Al-Jazeera T.V., July 21, 2006.

35 On increasing survivability, see for example: Nasrallah, March 19, 1999.

36 On the logic of using SSM and SSR, see for example: Hamas's senior commander, Al-Qassam website, February 1, 2009.

37 On the importance of inflicting pain and casualties, see for example: Saddam Hussein, speech on Baghdad domestic service, January 17, 1991.

38 On the role of the media, see for example: Al-Zawahiri, July 9, 2005; Abu Ubeida [and other senior members of Hamas's military wing], interview, Al-Jazeera TV, July 4, 2006.

39 On the logic of 'close battle', see for example: Saddam Hussein speech January 11, 1991; Nasrallah, speech, July 16, 2006.

40 On the importance of coping (passively) with air superiority, see for example: Nasrallah, March 19, 1999.
41 Bernard Lewis, *The Crisis of Islam: Holy War and Unholy Terror*, London: Weidenfeld and Nicolson, 2003.
42 On using Islam as a tool for mobilization and for increasing legitimacy, see, for example: Bin Laden, *Jihad against Jews and Crusaders.*
43 On using Islam as a mean of communication for explaining both the elements of O-RMA and the relevant tactical patterns, see for example: Nasrallah, February 19, 2009.
44 On using Islam as an intellectual source, see: S. K. Malik, *The Qur'anic Concept of War*, Adam Publishers and Distributors, Delhi, 1992.
45 On 'Jihad', see, for example: Hassan Al-Banna, 'Jihad' [from 1949] in Jim Lacey, *The Canons of Jihad: Terrorists' Strategy for Defeating America*, Naval Institute Press, Annapolis, pp. 4–10.
46 In 1995 one of Saddam's generals, Hamdany, offered to transform the Iraqi military to 'guerrilla like forces'. See Woods, *The Iraqi perspective report.*
47 Michael Eisenstadt and Kenneth M. Pollack, 'Armies of snow and armies of sand: the impact of Soviet military doctrine on Arab militaries' in *The Middle East Journal*, 55:4, pp. 549–70.
48 Donn A. Starry, 'The Legacy of Drummers, Warriors and Storytellers' in *Army Magazine* 52, July 2002.
49 Amos Gilboa, 'The Arab security strategies following the Yom Kippur War', [in Hebrew] in *Maarchot*, 361, November 1998.
50 On lessons from the first Lebanon War, see for example: Abu Ubeida, January 10, 2009; Bashar al-Assad, interview, *Asharq al-Awsat* [newspaper], June 12, 2000.
51 Ali Ahmad Jalali and Lester W. Grau, *Other side of the mountain: Mujahideen tactics in the Soviet–Afghan war*, 1995. On lessons from the war in Afghanistan as a source for O-RMA, see for example: Al-Zawahiri, *Knights Under the Prophet's Banner*; Bin Laden, *Jihad against Jews and Crusaders.*
52 On lessons from the Iran–Iraq War, see for example: Ali Shamkhani [Iranian Defense Minister], interview, *Siyasat e-Ruz*, February 18, 2003.
53 For a more comprehensive description of O-RMA by Nasrallah, see, for example: Nasrallah, February 22, 2008.
54 Quoted in: Matti Steinberg, *Facing their Fate*, [in Hebrew], Miskal, 2008.
55 On the Iraqi perspective toward the Second Gulf War, see Woods, *The Iraqi perspective report.*
56 On Hezbollah's preparation for the Second Lebanon War, see, for example: Andrew Exum, *Hezbollah at War: A Military Assessment*, Policy Focus No. 63, The Washington Institute for Near East Policy, December 2006.
57 On the debate concerning the outcome of the Second Lebanon War, see: Avi Kober, 'The Israel Defense Forces in the Second Lebanon War: Why the Poor Performance?' in *Journal of Strategic Studies* 31:1, February 2008, pp. 3–40; Edward Luttwak, 'Misreading the Lebanon War', *Jerusalem Post*, August 20, 2006.
58 Carmit Valensi, *The Emergence of Hybrid Actors*, [in Hebrew], Tel-Aviv University, 2011.
59 Frank Hoffman, 'Hybrid warfare and challenges' in *Joint Forces Quarterly* 51, 2009, pp. 34–40.
60 Hassan Nasrallah, speech, *Al-Manar*, August 18, 2007.

9 Improving in war

Military adaptation and the British in Helmand, 2006–2009

Theo Farrell[1]

How do militaries improve operational performance in war? Improvements may involve major organisational change. But, equally, they may involve only minor change to how militaries operate, or indeed no change at all (just better implementation of existing organisational routines). Up to now the scholarly literature has focused on the imperatives and processes of major military change, i.e. military innovation.[2] But innovation is neither a necessary nor a sufficient condition for victory in war. What victory almost certainly requires is for militaries to adapt to the operational environment and challenges they face, both when they first deploy and as the campaign evolves. This chapter develops a theory of military adaptation, which it applies to an analysis of the British campaign in Helmand from 2006–2009. It shows how British brigades adapted different ways of using combat power to try and defeat the Taliban from 2006–2007, and how from late 2007, British brigades have adapted a new population-centric approach that has focused more on influence operations and non-kinetic activities.

Military adaptation

It is entirely understandable that scholars in strategic studies should have concentrated on explaining military innovation.[3] Change on such a scale can have significant implications for national force structure and the defence budget, as well as on the balance of power and the offense–defence balance between states. By comparison, tinkering with tactics, techniques and technologies seems far less significant. And yet often it is such tinkering that aids victory in battle and contributes to success in war. Refinements to U.S. tactics, techniques and procedures (TTPs) for roadblocks in Iraq suggest the importance of small tactical changes to the success of a military campaign.[4]

In his seminal book, *Winning the Next War*, Stephen Rosen defines military innovation in terms of big change – i.e. a new way of fighting or a whole new combat arm. Rosen distinguishes between peacetime, wartime and technological innovation. Rosen argues that peacetime innovation involves a top-down campaign of military change led by a visionary military leader. He also presents wartime innovation as mostly involving top-down change to realign military operations and strategic goals. Finally, for Rosen, technological innovation centres on the management of risk and uncertainty both concerning enemy technologies, and the cost and potential of new technologies.[5] Rosen is not alone in concentrating on top-down change; most works on military innovation do so. Indeed the main debate in the literature is over rival models of top-down (military versus civilian led) innovation.[6] Rosen's typology, whilst analytically useful, disguises the reality of innovation that often stretches across periods of peace and war, and usually involves some new technology. Indeed, the peacetime innovations examined by Rosen – amphibious, carrier, and airmobile warfare – were all refined in war.[7] And today what we have seen with the U.S. military is a broad-based programme of force transformation which self-consciously involves technological, organisational and doctrinal innovation to develop capacities for network-centric and effects-based operations.[8]

Unlike Rosen, this chapter focuses on how militaries seek to win the 'current war'. For this, we need a theory able to explain bottom-up change by organisations at war.[9] In other words, we need a theory of military adaptation. Military adaptation is here defined as change to tactics, techniques or existing technologies to improve operational performance. In contrast, I understand military innovation to be a major change that is *institutionalised* in new doctrine, a new organisational structure and/or a new technology. Hence, whilst innovation can occur rapidly (i.e. revolutionary innovation), it more often occurs over a long period (i.e. evolutionary innovation).[10] Of course, adaptation may lead to innovation: new tactics or techniques may in time be captured in doctrine, or lead to a change to organisational structure or the acquisition of a new enabling technology.

My theory of adaptation recognises military organisations to be both rational and routine-bound. They are rational in that they monitor their performance and their environment, and respond to under-performance and environmental change. But they are bound by routines in *how* they measure performance and perceive their environment, and *how* they generate and select appropriate responses.[11] Routine is generically used here to mean organisational norms, rules and procedures, which are codified in organisational rulebooks, doctrine, and training, and frame organisational perception, decisions, and action.[12]

Military organisations can adapt in two ways to underperformance and environmental change. First, they can *exploit* core competencies in refining or modifying existing tactics, techniques and/or technologies. Second, they can *explore* new capacities by developing new modes and means of operations.[13] Where exploration involves doctrinal or structural change, or the acquisition of a brand new technology, it crosses the threshold into innovation. Exploration may involve adaptive change that is below the innovation threshold where, for instance, it involves new tactics or techniques developed in the battlefield but not institutionalised in doctrine or structural change. Similarly, exploration might involve new uses of existing technologies, or indeed the introduction of a mature technology that is new to the organisation, but itself hardly innovative.[14]

Rationality and routines lead militaries to favour exploitation over exploration. First, militaries must balance the need to master the tactics, techniques and technologies to sustain core war-fighting competencies. Without the repetition in training, exercises, and operations, these core competencies can quickly deteriorate with the risk of defeat in battle.[15] Yet such routine and repetition retard the inclination and ability to explore new ways of doing things.[16] Second, exploration may threaten existing ways of war in which militaries have heavily invested, and around which have developed sub-community interests and cultures.[17] Third, militaries have good reason to favour tried and tested ways of doing things, and to be sceptical of the newfangled ones, because war is so unforgiving of failure.[18] The concept of 'competency trap' nicely captures the problem: given sunken costs, organisational routines, and vested interests, militaries find it very difficult to abandon that which they have become good at.[19]

Eventually, however, organisations must explore new ways of operating or they risk being overtaken by competitors and environmental changes. So when do military organisations cease exploiting and start exploring? Most obviously, when they really have to. That is to say, when they are at war and appear to be losing. The military innovation literature highlights the importance of prospective defeat as a trigger of change. This raises the question of what constitutes prospective defeat? The answer is bound to be context specific, and involve a mix of material indicators (e.g. relative rates of force attrition, civilian losses, and economic and infrastructural damage) and perception (e.g. national morale as reported in the media and recorded in opinion polls). Since military adaptation involves bottom-up change, then it is reasonable to infer that battlefield indicators and military perceptions will be most significant in triggering a shift to exploration. The military innovation literature also highlights the role of military entrepreneurs who champion innovations from a position of authority within their organisation, or who do so outside the organisation and with the support of civilian leaders. These two triggers can work in synergy. The experience or prospect of defeat can create the permissive conditions for military innovation by discrediting old means and modes of operations.[20] When it comes to innovation there is evidence to suggest that entrepreneurial leadership is often not enough either; even military chiefs, who have the vision and authority to effect great military change, can be unsuccessful. The organisation must recognise that it has suffered a catastrophic failure, or at least be made to see that it risks doing so.[21] However, we may reasonably expect that adaptation, even explorative, is easier for a military to contemplate and undertake than innovation because it does not require *institutional* change. New tactics and techniques that are adopted may be readily abandoned when the need for them has passed. Thus, we may infer that the risk and scale of defeat may not have to be as stark for exploration as for innovation. Equally, entrepreneurs may not require the urgency of defeat to see their new ideas put into practice.

Organisation theory further suggests three *enabling* factors for exploration. First is poor organisational memory. Organisational routines are encoded and transmitted through organisational memory. For military organisations, memory is stored in doctrine, lessons-learned processes, and training. Good memory facilitates exploitation of core competencies; effective doctrine, lessons-learning, and training all contribute to the recall and appropriate application of core routines.[22] In contrast, poor memory facilitates exploration by reducing the efficiency with which organisations recover and transmit core competencies. Poor memory also reduces the tendency of organisations to prematurely abandon new alternatives. This is a natural tendency, since organisations are intolerant of failure and most new ideas fail. But some new ideas, with refinement (in content or execution), can prove effective. As two prominent organisation theorists note: 'If failures are remembered poorly ... an alternative that is initially unsuccessful can overcome its bad reputation.'[23] The second enabling factor is the degree of centralisation in the organisation. Centrally controlled organisations are efficient at exploitation because they can allocate resources as necessary to sustain a sufficient repertoire of competencies, and ensure adequate application of competencies in operations. Decentralised organisations, where authority and a degree of autonomy are delegated to component units, are less efficient at the management of core competencies. But they are also more sensitive to changes in their local environment. These two characteristics, less focused on centrally managed competencies and more focused on local conditions, make decentralised organisations more inclined to explore new alternatives.[24] The third factor is personnel turnover. Ideas travel with people. Organisations can lose old knowledge when people leave. At the same time people can bring fresh ideas and perspectives with them when they join organisations.[25] This effect is most pronounced at the top; hence struggling businesses often try to turn things around by replacing their chief executives.

The British campaign in Helmand, 2006–2009

The British military campaign in Helmand from mid-2006 to mid-2009 evolved from one centred on hard military power and directed at destroying the Taliban, to one focused on generating 'soft effects' and securing the civilian population. Over this period no less than six British task forces deployed to Helmand, each one based around a brigade (with augmented units) serving a six month tour. In the initial phase of the campaign, from May 2006 to September 2007, three successive British brigades engaged in major combat operations against the Taliban. These brigades struggled with a campaign that was under-resourced and under intense operational and political pressure. They each responded by exploiting their own core competences in combined arms warfare.[26] In September 2007 the fourth British task force arrived in Helmand with a new population-centric approach, which led it to explore and develop new capabilities. Thereafter, the British campaign entered a second phase where the main effort has involved consolidation of this new approach focusing on influence operations rather than combat operations, and directed towards securing and developing key areas of Helmand.

Exploiting old ways of war

When the North Atlantic Treaty Organisation decided to expand the International Security Assistance Force (ISAF) to southern Afghanistan, Britain agreed to take responsibility for Helmand. As it transpired this was to be the most challenging province to stabilise in Afghanistan, given the scale and reach of the narco-economy, the level of corruption in public life, the degree of local support for the Taliban, and the extent of underdevelopment especially in terms of infrastructure.[27] This new campaign was a stretch for a small army that had 8,500 troops deployed in Iraq in 2005–2006.[28] The British Army then stood at 101,000, with seven deployable brigades. In addition the Royal Marines also had a deployable brigade: 3 Commando Brigade.[29] The heavy and medium brigades were all committed to the Iraq campaign. Accordingly the Army's elite light brigade, 16 Air Assault Brigade, was selected to provide the first British task force for Helmand, with 3 Commando Brigade to follow in the second tour.

However, the Ministry of Defence (MoD) decided to cap the U.K. task force at 3,150 troops; in other words only a half brigade in strength. Its primary job was to protect the Provincial Reconstruction Team (PRT) and to create security in the key ground between and surrounding the provincial capital of Lashkar Gah and the main economic town of Gereskh. PRTs are the primary vehicles for stabilisation and reconstruction in Afghanistan (there are currently 25 PRTs in Afghanistan). The British were taking over from a small U.S. PRT in Helmand: the British PRT was staffed by a handful of civilians from the Stabilisation Unit, the Foreign Office, and the Department for International Trade. The U.K.'s Permanent Joint Headquarters (PJHQ), at Northwood, deployed an advanced planning team to Afghanistan in November 2005. The lack of a significant ISAF presence in Helmand meant that there was little overt military action by the Taliban at the time, and this appears to have led PJHQ to underestimate the threat facing the British task force.[30] Given this, PJHQ decided to deploy the task force in company-sized parcels. When initial elements of the British force encountered stiff resistance from the Taliban it was decided to accelerate the deployment and send an additional 1,500 troops.

However, these measures were too little, too late. It still took three months for the task force to deploy in full, and hence it lost important momentum in its break-in battle.[31]

Under pressure from the British Government, the corrupt governor of Helmand, Sher Mohammed Akhunzada, had been replaced by Afghan President Karzai with a technocrat, Mohammed Daoud. Daoud failed to understand that the task force contained only one battle group of deployable infantry (namely 3 PARA battalion). Accordingly, he put pressure on the British to deploy forces to garrison towns in northern Helmand that were vulnerable to Taliban attack. The British task force commander appreciated the political imperative to support Daoud in this situation. Thus, the small British task force came to be stretched very thinly under a 'platoon house' strategy that saw the British garrison the towns of Sangin, Musa Qaleh and Now Zad. Moreover, the British garrisons promptly came under siege from Taliban forces, and British forces found themselves locked in a series of major battles. In the end, Now Zad and Musa Qaleh became so difficult to resupply (requiring very dangerous helicopter missions), that 16 Brigade eventually had to abandon these towns. Woefully under-resourced, the British task force truly found itself between a rock and a hard place. British commanders had little choice but to accede to Daoud's requests for military assistance, especially as they came under pressure from Karzai and from Whitehall to do so. At the same time they came under fire from the U.S. head of Regional Command (South) (RC(S)), Major General Ben Freakley, because with practically the entire British task force defending towns from Taliban attack, the British could contribute little to support the RC (South) scheme of manoeuvre.[32] In its post operation report 16 Brigade argued that the platoon house strategy had 'undoubtedly contributed to the attrition of Taliban forces'. However, 16 Brigade also recognised that since 'the focus of 3 PARA battle group's activity was on the delivery of kinetic effect [...] the availability of assets to support non-kinetic and PRT activities was therefore limited'.[33]

In October 2006, 3 Commando Brigade took over from 16 Brigade with a very different concept of operations (CONOPS). The Royal Marines were also focused on the economic and political centre of Helmand, which was designated an Afghan Development Zone (ADZ). But the Marines adopted a very different approach, one which ironically took the focus of their activities outside the ADZ. The problems had been observed by 3 Brigade with the platoon house strategy, which had fixed British forces and prevented them from supporting development activities. The new task force commander, Brigadier Jerry Thomas, declared 'reconstruction of the ADZ' to be his main military effort. However, given the dire security situation, Brigadier Thomas determined that 'security represents the current pre-eminent line of activity' and that 'once our security is in place, development will gather momentum and provide the best tool with which to counter this insurgency'.[34] For the security line of operation,

Thomas intended to 'unfix the North' and 'manoeuvre to threaten, disrupt and interdict the enemy'.[35] To this end, 3 Brigade created a number of Mobile Operations Groups (MOGs) – 250-strong flying columns in 40 vehicles (a mix of Vikings and Land Rovers) – which were tasked with seeking out and engaging the Taliban. Here the Royal Marines were drawing on their core competency in conceiving of the 'desert as a sea', through which MOGs could roam wide and hunt down the enemy. However, the Taliban proved too wily to be drawn by the MOGs into prepared kill zones and often as not the MOGs ended up engaging the Taliban on their terms. Indeed, this led Royal Marines to coin the term 'advance to ambush' to describe MOG operations.[36]

3 Brigade went into Helmand with the right intent. Influence was to be at the centre of their CONOPS. The brigade intended to 'tread softly', only escalating to violence as required and with due care to minimise collateral damage so as to 'avoid breeding new enemies'. 3 Brigade's Operational Design correctly noted 'almost all tactical engagements (win or lose) favour the enemy. By definition, they demonstrate instability and insecurity thus undermining perceptions of Government of Afghanistan influence, control, and credibility'. However, fighting in Helmand *escalated* under 3 Brigade. The total number of engagements with enemy forces went up from 537 under 16 Brigade's tour, to 821 under 3 Brigade's tour.[37]

In its post operation report, 3 Brigade claims to have thrown the Taliban off balance: 'The "dynamic unpredictability" arising from being able to engage the enemy at times and places of our choosing has disrupted him, undermined his will and shattered his unity.' It is clear that the Royal Marines were more successful at taking the fight to the enemy than 16 Brigade had been: under 3 Brigade there was a fourfold increase in the number of engagements initiated by British forces (up to almost 300 engagements). However, as noted earlier, evidence from the field suggests that many of these engagements were actually at a place of the enemy's choosing. Moreover, the brigade that followed found Taliban will and unity to be far from 'shattered'. The larger point is that 3 Brigade never managed to create the security necessary to allow stabilisation and development to proceed. The MOGs mostly operated outside the ADZ. This had some effect in keeping the Taliban at bay, resulting in a 45 per cent reduction in attacks on British bases guarding district centres. But it also reduced the presence, and hence enduring security effect, of 3 Brigade within the ADZ.[38]

Both 16 and 3 Brigades are specialist, high readiness brigades. With the deployment of 12 Mechanised Brigade in April 2007, Britain committed one of its regular line brigades. 12 Brigade realised that the mobile operations of 3 Commando Brigade were doing little to secure the populated areas of Helmand. 12 Brigade's observation was that the Royal Marine MOGs 'would manoeuvre around in the desert, hit the crust [i.e., the edge of the Green Zone[39]], have a contact [with the Taliban] at range, but have no enduring effect, rarely penetrating into the Green Zone'.[40] Unlike 3 Brigade, 12 Brigade focused on securing the ground around and between population centres inside the Green Zone, especially the area between Gereshk (the main economic centre of Helmand) and the newly captured town of Sangin. This involved a series of major 'clearance operations', leading to a further escalation in fighting with the number of engagements with enemy forces rising from 821 under 3 Brigade to 1,096 under 12 Brigade.[41] The Task Force commander, Brigadier John Lorimer, was prepared for an increase in fighting on 12 Brigade's tour. The Taliban were promising a summer offensive and so Lorimer was determined that if there was to be an offensive, that it would be an ISAF one.[42] Hence, he emphasised core combat skills to be 'mastered by everyone, irrespective of rank or cap badge' in pre-deployment training for his brigade.[43]

At the heart of 12 Brigade's CONOPS was creating security through 'persistent presence'. Brigadier Lorimer appreciated that security for Afghans meant 'compound walls'; i.e. solid evidence of security force presence. It was also well understood that areas cleared of Taliban had to be held in order to prevent the insurgents from simply returning. Accordingly, 12 Brigade established a number of patrol bases in the Upper Gereshk Valley. The hope was to backfill areas cleared by British forces with Afghan National Army (ANA) and Afghan National Police (ANP) units. However, much like its predecessors, 12 Brigade simply lacked the resources to realise its ambitions. For most of his tour Brigadier Lorimer only had a battle group and some rifle companies of deployable forces. Most of the additional British forces that came into theatre with 12 Brigade were actually 'enablers' (i.e. Combat Support and Combat Service Support). The ANA were unable to fill the gap in security forces. There were three ANA kandaks (battalions) in Helmand; one each in Gereshk and Sangin, and one for training and force generation at Camp Shorabak. These were perfectly fine at offensive operations, provided they received sufficient mentoring and logistical support from TFH, but they were poor at holding ground. The ANA view was that this was a job for the police. The ANA preferred to return to barracks once offensive operations had been concluded. The ANP were extremely poor quality in Helmand, riddled with tribal factionalism, outright corruption and drug abuse. Indeed they were far more of a destabilising factor than a force for stability. Despite their best efforts, 12 Brigade lacked the forces and Afghan partners to provide for enduring presence in territory cleared of Taliban.[44] Accordingly the Taliban were able to re-infilitrate. The lack of real progress led a frustrated Brigadier Lorimer to reflect that it felt rather like 'mowing the lawn'.[45]

Exploring a new way of war

The military campaign changed direction with 52 Infantry Brigade in October 2007. The incoming brigade's CONOPS was 'clear, hold, build'.

In contrast, 12 Brigade had done little actual holding and no building. Moreover, 52 Brigade's Operational Design conceptualised the campaign centre of gravity in terms of the local population instead of the enemy's will and ability to fight. Indeed, for the new task force commander, Brigadier Andrew Mackay, the Taliban body count was a 'corrupt measure of success'.[46] Accordingly, 52 Brigade's campaign focused on influence operations to win the consent of the population. Hence, whereas 12 Brigade rotated units through its FOBs, 52 Brigade committed units to FOBs for the entire tour in order to 'provide reassurance and provide clear proof that our presence is not transient or temporary'.[47] For the same reason 52 Brigade also dismantled one FOB that was in an unpopulated area, and built two new FOBs inside the Green Zone.[48] As the brigade plans officer noted, the position of FOBs was important 'to establish[ing] a pattern of life locally'. He also observed how 'locals had something of a "castle of the hill" mentality' and that by having an enduring presence, 'locals would begin to engage the British and provide information'.[49]

Thus, influence operations were absolutely central to 52 Brigade's campaign. Previous brigades had not ignored influence operations. But they had not given them the same prominence in their operational design, nor resourced them to the same level as 52 Brigade. 52 Brigade had more officers working on influence operations than its predecessors in task force HQ, and Brigadier Mackay also appointed dedicated influence officers in each of his battle group HQs.[50] Furthermore, 52 Brigade developed new capabilities for influence operations. Two new capabilities stand out. One was the creation of teams that were tasked with helping to develop, deliver and assess non-kinetic effects and influence operations; this included two-man Non-Kinetic Effects Teams (NKETs) which were deployed down to company level, and larger (up to 20 strong) Development and Influence Teams (DITs) which provided a deployable task force asset.[51] The second new capability was the Tactical Conflict Assessment Framework (TCAF). This was a diagnostic methodology to identify the causes of instability in a unit's area of operations. It essentially involved patrols asking local people a set of four questions about the issues that most concerned them.[52] Multiple and iterative questionnaires were used to identify local-level trends and priorities, and this data was used to target appropriate development and stabilisation activities. Follow-up TCAF questionnaires were used to measure the effectiveness of these activities.[53] As one officer observed, TCAF 'was not a non-kinetic silver bullet'. A trial of TCAF in Lashkar Gah appeared to show very positive results in terms of gathering intelligence to target non-kinetic activities. However, this may have created unrealistic expectations in the task force headquarters, and it would appear that TCAF was rolled out too quickly across Helmand, and especially into districts where the environment was not conducive to regular interaction with non-hostile local populations.[54] 52 Brigade may have pushed TCAF too far and too fast, but it did at least invest in developing this new capability for influence operations.

The contrasting approaches of 52 and previous brigades is also evident in their respective approaches to battle. Clearance operations by earlier brigades had involved such extensive use of the force that they often caused the population to flee their homes, in many cases never to return. When 52 Brigade was tasked with retaking Musa Qaleh from the Taliban, Brigadier Mackay was determined that this should be done in a way that minimised the displacement of the local population. For Brigadier Mackay, 'the people were the prize', not ground or enemy attrition. So he planned a cautious offensive, involving armoured forces slowly approaching the town on both flanks weeks in advance of the main assault. The intention was not to capture and kill Taliban forces, but rather to coerce them into leaving the town and thereby avoiding the necessity for a major battle. This approach also gave the local population plenty of warning so they could seek safety in advance of the final assault. This contrasted with the major clearance operations by 12 Brigade, most of which proceeded with no prior messaging to warn the local population because it did not want to alert the Taliban.[55] In contrast, 52 Brigade was prepared to compromise operational security in order to keep the local population onside.[56] The plan largely worked; the Taliban fled Musa Qaleh after a brief fight.[57] Within hours of recapturing the town, 52 Brigade started preplanned stabilisation activities. As one journalist on the ground noted: 'The residents of Musa Qala voted with their feet soon after the arrival of the Afghan army. Tractors, pick-up trucks and carts started bringing people home.'[58]

This new population-centric approach continued under 16 Brigade on its second tour in Helmand from April to October 2008. This time under the command of Brigadier Mark Carleton-Smith, the brigade's CONOPS was to 'go deep not broad'. Hence the British Task Force focused on protecting population centres and on developing the Afghan government's influence and authority in those areas that realistically could be secured and held. In terms of the enemy, the focus was on undermining Taliban influence rather than fighting its forces. Like 52 Brigade therefore, 16 Brigade strove to achieve an appropriate mix of kinetic and non-kinetic activities. Indeed, Brigadier Carleton-Smith prepared his brigade for a very different campaign than the one conducted by 16 Brigade in 2006: 'I wanted to dispel any danger that 16 Brigade might have confused the intensity of the fighting on HERRICK 4 with the nature of the campaign we were seeking to prosecute.'[59] Arriving in theatre, 16 Brigade maintained almost all the innovations of its predecessor; e.g. NKETs, inclusion of civilians in the Joint Targeting Board, and partnering of ANA battalions with task force battle groups. An exception was TCAF, which 16 Brigade abandoned in August 2008 because it judged it to be unreliable and of limited utility.[60]

Critically, like Mackay, Brigadier Carleton-Smith emphasised the centrality of influence operations over combat operations:

> Our plan for the Taliban was to try to undermine their strategy, rather than merely fight their forces. Their strategy seemed to be more one of influence, intimidation, and the provision of parallel shadow government than of tactical engagement. Therefore the aim was to marginalize their influence in the centres of population.[61]

Thus brigade command staff understood that influence is 'everything we do'.[62] To this end 16 Brigade developed greater integration with the expanding British-led PRT in the provincial capital, Lashkar Gah. The brigade planning section, J-5, was physically moved to the PRT HQ. Moreover, the deputy brigade commander, Colonel Neil Hutton, became one of the three PRT deputies. Under 16 Brigade the PRT more than doubled in size: the civilian staff grew from 45 to 94, and military personnel from 19 to 48. By the end of its tour 16 Brigade concluded that the PRT had 'a structure and a process' that would enable it to deliver the British interagency plan for Helmand.[63]

Explaining British military adaptation

The British military campaign in Helmand has been a highly adaptive one. Early innovation was not on the cards: defence chiefs and civilian policymakers did not fear outright defeat in Helmand.[64] Nonetheless, we see from post-tour reports that successive commanders perceived the campaign to be failing, and this drove each to take a different approach to his predecessor. Under 3 Commando Brigade and 12 Mechanised Brigade in 2006–2007, adaptation involved exploitation of core competencies. 3 Brigade sought to avoid the mistakes of 16 Brigade by drawing on a very familiar form of warfare involving MOGs. In turn, 12 Brigade tried to avoid the mistakes of 3 Brigade by clearing ground, but the mechanised approach it took was not conducive to holding ground, let alone securing the population. In contrast, with 52 Brigade we see exploration of a new population-centric, less kinetic approach focused on influence operations. Why did the British way of war in Helmand change in late 2007 and not before?

Enabling exploration

The British military had, and still has, all the characteristics that enable exploration. It has weak organisational memory. It invests relatively little in developing and promulgating lessons learned. The MoD's lessons management system does generate 'lessons' and requires a response from the relevant defence 'stakeholders'. But this is a mechanical process that encourages a tick box approach and fails to distinguish and prioritise big operational lessons from small technical lessons.[65] Traditionally the British military has also been largely disinterested in formal doctrine, relying instead on informal officer networks to transfer corporate knowledge.[66] In the 1980s the British Army came to appreciate the need to get serious about doctrine and so, from the 1990s onwards, the British became significant producers of military doctrine.[67] However, the view within the British Army going into Helmand was that it lacked an up-to-date COIN doctrine. Thus 52 Brigade, looking for doctrine to inform its training and preparations, drew on the new U.S. Army/U.S. Marine Corps doctrine on COIN – Field Manual (FM) 3–24 – which was issued to all commanders in the brigade. Indeed, 52 Brigade's CONOPS, 'clear, hold, build', was taken directly from FM 3–24.[68] Actually, the British Army did have a COIN doctrine, produced in 1995, that was fit for purpose: and indeed, this doctrine informed FM 3–24. However, *Countering Insurgent Operations* (1995) was not mass produced and promulgated because the British Army's main focus in the mid-1990s was on peace operations, and so was little-known-about a decade later.[69] In other words poor organisational memory prevented the British Army from recovering core competencies. At the same time, as the case of 52 Brigade illustrates, it gave impetus to the search for new ideas.

The British military campaign is fairly decentralised, insofar as the task force commander is given considerable leeway to shape his own campaign. Formally, the task force commander works within two chains of command. There is an ISAF command chain that runs from the Commander ISAF (COMISAF) to the Commander of RC(S), down to the British task force commander. There is also a national command chain that runs from the British National Command authorities through PJHQ to the British task force commander. Successive post-tour reports indicate that British task force commanders have never felt compelled to follow directives from COMISAF. Reinforcing this is the British military culture of command (called 'mission command') that seeks to preserve the autonomy of the commander. Hence, PJHQ gives an overall guidance but leaves it up to the task force commander to define his own CONOPS and operational design. 52 Brigade was operating to the Joint Plan for Helmand (which was produced in late 2005 and early 2006). But this joint plan was pretty thin on detail, especially in the military line of operations. The plans officer for 52 Brigade noted how the brigade 'received very little in the way of higher formation direction in campaign terms' and hence 'we were at liberty to identity what effect we wanted to have, where, and (to a greater or lesser degree) when we wanted it to be put into effect'.[70] As it transpired the Joint Plan for Helmand was replaced during 52's tour by a new inter-agency plan called the Helmand Road Map (HRM). More to the point, the HRM was produced by the PRT and 52 Brigade planners in theatre, and was subsequently adopted by the U.K. Government.[71]

Staffing of campaigns by the British military involves regular personnel turnover. As noted, the British task force rotates every six months and with it so do the British commander and his staff. This contrasts with longer 9–12 month tours of U.S. brigade combat teams in Iraq and Afghanistan, and even longer tours of U.S. command staffs (which can last two years). The lack of continuity of command can obviously cause problems, in terms of taking the longer perspective. But it has provided regular opportunities for the British task force to adopt a different approach in response to a failing campaign, as attempted by three brigades (3, 12, 52) between 2006 and 2008.

The puzzle remains, however. Since poor organisational memory, decentralisation, and personnel change were present from the beginning of the campaign, why did it take 18 months for the British military to shift from exploiting core competencies to exploring a new approach? Here the context of the campaign mattered.

A more favourable context

52 Brigade did benefit from certain contextual factors that facilitated the shift to a 'softer' way of war, namely new Taliban tactics, increased ANA capabilities and increased resources. In 2006–2007, Taliban forces suffered considerable attrition in heavy fighting with ISAF in Helmand. British Defence Intelligence puts the number of Taliban dead in the thousands (though some British commanders have expressed doubts about such high figures).[72] Accordingly, since early 2008, the Taliban have been less inclined to launch major assaults on district centres and ISAF bases. Thus, when U.S. Marines launched an offensive against the Taliban strongholds in Garmsir district in 2008, the Taliban main force retreated rather than put up a fight. Equally, when 16 Brigade launched an air assault on Taliban villages south of Musa Qaleh, they found that the Taliban had fled.[73] In Kajaki an Afghan interpreter hired by the British to listen to Taliban communications in 2008 'described almost comical attempts by different commanders to shirk combat and foist the responsibility on other commanders'.[74] Essentially the Taliban learned the cost of engaging in direct attacks on ISAF forces. Instead they have made increasing use of IEDs to fix British forces and to defend their own positions. Most ISAF casualties are now caused by Improvised Explosive Devices (IEDs).[75] The Taliban did continue to engage in firefights with British forces. Hence, at other times, 16 Brigade engaged in some fierce firefights with Taliban around Musa Qaleh.[76] And when 3 Commando Brigade rotated into Lashkar Gah, on their second tour in October 2008 to replace 16 Brigade, they discovered a large Taliban force (some 300 strong) moving into position to attack the provincial capital: this force was destroyed by ISAF air power.[77] Overall, however, we see a more cautious approach from 2008 by a battered Taliban which, whilst no less deadly to ISAF forces, has been

more susceptible to stabilisation operations because it has resulted in fewer battles in and around district centres.

Just as the Taliban were becoming more cautious in 2008, so ANA capabilities were improving. The ANA grew from around 50,000 in 2007 to almost 80,000 in 2008.[78] Moreover, the ANA increased its ability to conduct battalion-level operations; 30 per cent of ANA battalions could do so with the support of international forces by December 2006, rising to 44 per cent by December 2007 and 62 per cent by December 2008.[79] The situation in Helmand reflected this national trend. Accordingly the British task forces have been increasingly able to integrate ANA into their campaign. Indeed, the mentoring of ANA and ANP was another central element of 52 Brigade's CONOPS which gave prominence to 'a determination to facilitate, coordinate and encourage greater Afghan ownership of the security architecture within HMD', and recognised that this would 'involve risk and will require patience and steady nerves but we must be prepared to follow Afghan advice to Afghan problems and be part of an Afghan-led solution'.[80] By mid 2007 there were five ANA battalions deployed in Helmand, with a sixth battalion added in late 2008. Over this period we see a progression from using the ANA to backfill areas secured by the British to (from late 2007 on) the partnering of ANA battalions with British battle groups and the ANA being given its own independent area of operations within Helmand.[81]

Finally, the campaign has received ever more support and resources, especially as the British Government shifted its strategic focus from Iraq to Afghanistan. Hence, British troop numbers more than doubled between late 2006 and late 2007, peaking in 2008 at just over 8,500 (see Table 9.1). It should be noted that not all these troops were assigned to the task force in Helmand: a couple of thousand (including one battle group) were assigned to supporting RC(S) in Kandahar. At the same time the British task force was bolstered by an Estonian company and a Danish battle group (deployed in the Centre of Helmand); the U.K. Task Force was

Table 9.1 British troop numbers in Afghanistan

Task force brigade	*Date*	*Size*
16 Air Assault Bde.	April–October 2006	3,150 (4,500*)
3 Commando Bde.	October 2006–April 2007	5,200
12 Mechanised Bde.	April–October 2007	6,500
52 Bde.	October 2007–April 2008	7,750
16 Air Assault Bde.	April–October 2008	8,530
3 Commando Bde.	October 2008–April 2009	8,300

Source: Data from ISAF Troops Placemat arch.: www.nato.int/isaf/docu/epub/pdf/placemat.html.

Note

* British numbers rapidly grew to 4,500 with reinforcements to the Helmand Task Force.

renamed Task Force Helmand (TFH) in 2008 to reflect the inclusion of international partners. In addition TFH received support from various U.S. units operating in Helmand over 2007–2008. Equally significant has been the improved equipment for British forces. Better armoured vehicles – new Mastiffs, the British MRAP (Mine-Resistant Ambush Protected), and an Armoured Infantry Company in Warriors – arrived by the end of 12 Brigade's tour and so this gave crucial additional capability for subsequent task forces.[82] There was also more artillery with the deployment of two GMLRS (Guided Multiple Launch Rocket System) batteries; so accurate is the fire from GMLRS that it is called 'the 70 kilometre sniper rifle' in Helmand. That said, the key equipment shortfall, noted by all British commanders in their post-tour reports, was the shortage of helicopters and this problem persisted well into 2009.

In sum the operational landscape changed in Helmand between 2006 and 2007, with a shift in Taliban tactics, rising ANA numbers and capability, and increased resources available to the British-led task force. This resulted in a far more permissive context in which the British could develop a more influence-orientated and population-centric approach. Arguably these favourable factors were already present under 12 Mechanised Brigade. The final missing piece of the jigsaw is the character of 52 Brigade and its commander. As a long-standing regular brigade (formed in 1899), 12 Mechanised Brigade had an extensive and well-developed and exercised repertoire of conventional combat competencies that had been employed in the 1991 Gulf War and in Iraq in 2004. In contrast, 52 Infantry Brigade was formed for the First World War and disbanded afterwards, then reformed for the Second World War and again disbanded afterwards. In the 1960s it reformed but only as a reserve brigade. In 2002 it took command of regular army units but as a Type B brigade, i.e. a non-deployable regional brigade. In February 2006 it was turned into a Type A deployable brigade. In late 2006 Brigadier Mackay was notified that his brigade was to be deployed to Helmand within a year. 52 Brigade staff then underwent a massive expansion; the brigade HQ increased from 15 to (eventually) 175 staff.[83] Incoming staff officers brought a wealth of experience with them. But the key point is that this was a newly formed Type A brigade, and hence one that was less committed to an established repertoire of core competences and one more open to new alternatives introduced by Brigadier Mackay. Moreover, Brigadier Mackay was less conventionally-minded that his predecessors in Helmand. Prior to taking over in Afghanistan he had led missions to reform the Iraqi Police Force (2004) and the Lebanese Army (2006).[84] On his return from Helmand Brigadier Mackay was to work in the Development, Concepts and Doctrine Centre promoting the influence-orientated approach to operations.[85]

Conclusion

The British military has developed new tactics and capabilities for a more population-centric approach to its campaign in Helmand. To be sure, the British task force still relies on hard combat power to defeat Taliban field forces. It still conducts major sweeps to clear Taliban out of key areas. Hence, over June and July 2009, the TFH under 19 Light Brigade conducted a major offensive against Taliban strongholds in Babaji (which lies in the Green Zone between Lashkar Gah and Gereshk). Operation PANCHAI PALANG involved 3,000 British troops as well as Danish and ANA forces: ten British troops were killed in the operation along with many hundreds of Taliban.[86]

Still, there is change in the British approach that can be witnessed in the emerging emphasis on influence operations and non-kinetic effects, on establishing an enduring presence, and on developing the civil–military partnership. This change started with 52 Brigade, which developed new capabilities for influence operations. It was consolidated by 16 Brigade (on its second tour), which embedded its military planners in the civilian-led PRT. The new approach was continued under 3 Brigade (also on its second tour) and under 19 Brigade. 3 Commando Brigade built on civil–military integration by developing a Joint Command Board for the PRT and TFH, which was chaired by the PRT civilian head.[87] 19 Brigade also placed an emphasis on district stabilisation teams (comprising civilian and advisors and military support officers), which it saw as 'pivotal' to the British civil–military campaign, especially in stabilising areas 'on the heels of clearance operations'.[88]

This chapter has shown that successive British brigades have responded to the operational challenges in Helmand by adapting. The first brigade to deploy into Helmand had an appalling time of it. Woefully under-resourced and under immense pressure from different quarters to overextend, 16 Air Assault Brigade found itself fixed in several garrisons spread out across Helmand. The two brigades that followed each adapted to the mistakes of the last by exploiting their own core competencies. 3 Commando Brigade adopted a highly mobile approach to unfix the British task force. This failed to provide enduring presence, especially in the ADZ. 12 Mechanised Brigade conducted a number of large clearance operations to clear the Taliban out of the ADZ but these proved ineffective when it came to holding ground afterwards. The fourth brigade to deploy adapted by exploring new ways of operating. 52 Infantry Brigade put influence operations at the centre of its CONOPS and developed new capabilities for non-kinetic activities. As noted above, this new population-centric approach was consolidated by the brigades that followed: 16 and 3 (both on their second tours), and 19 Brigade.[89]

Whilst the academic literature is dominated by top-down accounts of military innovation (involving a mix of technological, structural and doctrinal change), the British campaign in Helmand confirms the importance of studying bottom-up military adaptation.[90] Most of the time, when encountering problems in the field, militaries will adapt. And oft-times, such adaptations are enough to turn a campaign around. Adaptation may involve better exploitation of existing tactics, techniques and technologies, or it may involve exploring new ways and means of operating. This was the case with the British in Helmand, with the development of a population-centric approach and new capabilities for influence operations and for generating soft effects.

Over time exploration may accumulate into what might be properly called military innovation. Arguably this is occurring with the British military in Helmand. The population-centric approach focused on influence operations is being institutionalised in structural and doctrinal changes. On the recommendation of 3 Brigade in 2008, 19 Brigade created, and 11 Brigade (which followed on) have retained, an Influence Cell in the task force HQ led by an SO1 (Colonel) Influence; under previous brigades, 'influence' was seen as the responsibility of psychological operations or media operations staff.[91] Furthermore the British Army has produced a new field manual on COIN which recognises that influence 'underpins everything which British forces undertake because counterinsurgency is as much about the battle of perceptions as it is about military operations targeted against insurgents'.[92] The MoD is also producing cultural advisors to be deployed in Helmand. Cultural understanding is essential for an influence-orientated, population-centric approach.[93] The push for this capability came from the British Army's Land Headquarters and from the Directorate of Joint Capability in the MoD in 2008, both of which identified the need for cultural specialists to support the Afghanistan campaign. In February 2009 an Afghan Specialists Joint Implementation Team was established in the MoD to develop this capability. By October 2009 a Cultural Advisor and a Cultural Coordinator were deployed to the Influence Cell in TFH. Cultural Advisors will be deployed in each of the British battlegroups by February 2010. In addition, basic training and pre-deployment training is being adapted to develop cultural awareness in military personnel.[94] The development of these new cultural capabilities has also been supported by new doctrine promulgated in January 2009.[95]

It remains to be seen how long lasting this innovation in influence operations will be. History suggests that, time and again, the British Army has forgotten innovations from COIN campaigns (innovations often institutionalised in campaign-specific doctrine) and has had to relearn them. Indeed, contrary to the picture offered by John Nagl of the British Army as a 'learning organisation',[96] it would appear that the British Army went through a series of campaign-specific learning cycles in the twentieth century, in each case to correct an approach that was overly military and kinetic in approach, and develop a more comprehensive and (what would today be called) population-centric approach.[97] In this respect the three organisational attributes that assist exploration – poor memory, decentralisation and personnel change – may well hinder the process of innovation, specifically the institutionalisation and recovery of innovations.[98]

This chapter has not explicitly discussed the question of military learning. Obviously organisational adaptation involves learning, both from contemporary experience and from the organisation's past. This, in turn, raises important questions about the specific modalities of British military learning. How did field brigades learn, both in preparing for and during operational deployment? What learning occurred at the institutional level, i.e. the British Army, Royal Marines, and joint institutions (especially PJHQ)? What learning occurred across theatres; in other words, what lessons did the British take from Iraq to Afghanistan? Finally, to what extent did the British learn from other militaries, especially from the U.S. military? The example of 52 Brigade drawing on the new U.S. COIN field manual (FM 3–24) in developing its CONOPS suggests that the British were learning from the Americans. However, the picture is complicated because, as noted earlier, FM 3–24 itself drew on British COIN doctrine. True learning, much like explorative adaptation, is not something that comes easily to the military.[99] But the experience of Afghanistan suggests that field formations do learn, and that learning is integral to military adaptation. More research needs to be done to understand precisely how learning occurs. This chapter has sought to illuminate the end point of learning, namely how militaries adapt. It has also shown just how much the British have adapted in Helmand.

Notes

1 Drafts of this chapter were presented at a conference on 'Military Theory in the Post-Cold War Era' hosted by the Norwegian Institute for Defence Studies in Oslo 24–25 June 2009, the Oxford University Strategic Studies Group at All Soul's College on 19 January 2010, the Defence Studies Seminar at the U.K. Joint Services Command and Staff College on 9 March, the Carr Center at the University of Harvard on 29 March 2010, and the Security Studies Program at MIT on 31 March 2010. The author is grateful to the participants of these events for their feedback. He also thanks the following for their written comments on earlier drafts: Huw Bennett, Frans Osinga, Peter Viggo Jakobsen, Mikkel Vedby Rasmussen, Terry Terriff, and Peter Dahl Thruelsen. The work on this chapter was made possible by an ESRC/AHRC Research Fellowship (RES-071–027–0069) funded under the Research Council's U.K. 'Global Uncertainties' programme.

2 Theo Farrell and Terry Terriff (eds), *The Sources of Military Change*, Lynne Rienner, 2002; Adam Grissom, 'The Future of Military Innovation Studies' in *Journal of Strategic Studies* 29:5, 2006, pp. 905–34.

3 There are numerous studies by historians of tactical change, especially by armies in the field. But historians have yet to suggest a theory of military adaptation. See, for example, Paddy Griffiths, *Battle Tactics on the Western Front: The British Army's Art of Attack, 1916–18*, Yale, NH: Yale University Press, 1994; and Michael D. Doubler, *Closing with the Enemy: How GIs Fought the War in Europe, 1944–1945*, Lawrence: University Press of Kansas, 1994.

4 Colin H. Kahl, 'In the Crossfire or the Crosshairs? Norms, Civilian Casualties, and U.S. Conduct in Iraq' in *International Security* 32:1, 2007, pp. 27–8.
5 Stephen Peter Rosen, *Winning the Next War: Innovation and the Modern Military*, Ithaca, NY: Cornell University Press, 1991.
6 The rival models are provided by Stephen Rosen (military led) and Barry Posen (civilian led). Rosen, *Winning the Next War*; Barry R. Posen, *The Sources of Military Doctrine: France, Britain, and Germany Between the World Wars*, Ithaca, NY: Cornell University Press, 1984. For studies which provide top-down models that emphasise civil–military interaction, see Deborah D. Avant, *Political Institutions and Military Change: Lessons from Peripheral Wars*, Ithaca, NY: Cornell University Press, 1994; Elizabeth Kier, *Imagining War: French and British Military Doctrine Between the Wars*, Princeton University Press, 1997; Peter Trubowitz, Emily O. Goldman and Edward Rhodes (eds), *The Politics of Strategic Adjustment: Ideas, Institutions, and Interests*, Columbia University Press, 1999.
7 See chapters on these innovations in Williamson Murray and Allan R. Millet (eds), *Military Innovation in the Interwar Period*, Cambridge University Press, 1996.
8 Frans Osinga, 'Military Transformation: an Introduction' in Terry Terriff *et al.* (eds), *A Transformation Gap? American Innovations and European Military Change*, Stanford, CA: Stanford University Press, 2010.
9 This is noted in Eliot A. Cohen, 'Change and Transformation in Military Affairs' in *Journal of Strategic Studies* 27:3, 2004, pp. 400–1; Grissom, 'The Future of Military Innovation Studies', pp. 920–4.
10 Williamson Murray, 'Innovation: Past and Future' in Murray and Millet (eds), *Military Innovation*, pp. 306–10.
11 W. Richard Scott, 'Symbols and Organizations: From Barnard to the Institutionalists' in Oliver E. Wiliamson (ed.), *Organization Theory: From Chester Barnard to the Present and Beyond*, Oxford University Press, 1995, pp. 38–55.
12 For studies on how routines frame the perceptions, decisions and actions of military organisations, see Lynn Eden, *Whole World on Fire: Organizations, Knowledge, and Nuclear Weapons Devastation*, Cornell University Press, 2004; Isabel V. Hull, *Absolute Destruction: Military Culture and Practices of War in Imperial Germany*, Ithaca, NY: Cornell University Press, 2005.
13 James G. March, 'Introduction', in March (ed.), *The Pursuit of Organizational Intelligence*, Blackwell, 1999, p. 5.
14 On the enduring, everyday importance of 'old' technology, see David Edgerton, *The Shock of the Old: Technology and Global History since 1900*, Profile Books, 2006.
15 For a recent articulation of this argument, see Gian P. Gentile, 'Learning, Adapting and the Perils of the New Counter-insurgency' in *Survival*, 51:6, 2010, 189–92.
16 Barbara Levitt and James G. March, 'Chester I. Barnard and the Intelligence of Learning' in Olliver Williamson (ed.), *Organization Theory*, 28.
17 Carl H. Builder, *The Masks of War: American Military Styles in Strategy and Analysis*, Baltimore: Johns Hopkins University Press, 1989.
18 Organisational mistakes and catastrophic military failure is explored in Eliot A. Cohen and John Gooch, *Military Misfortunes: The Anatomy of Failure in War*, New York: The Free Press, 1990.
19 James G. March and Barbara Levitt, 'Organizational Learning' in March (ed.), *Pursuit*, pp. 78–9.
20 Posen, *Sources of Military Doctrine*; Emily O. Goldman, 'International Competition and Military Effectiveness: Naval Air Power, 1919–1945' in Risa Brooks and Elizabeth A. Stanley (eds.), *Creating Military Power: The Sources of Military* Effectiveness, Stanford University Press, 2007, pp. 158–85; Joao Resende-Santos, *Neorealism, States and the Modern Mass Army*, Cambridge University Press, 2007.

21 Terry Terriff, 'Warriors and Innovators: Military Change and Organizational Culture in the U.S. Marine Corps' in *Defence Studies* 6:2, 2006, pp. 215–47.
22 March and Levitt, 'Organizational Learning', pp. 83–6.
23 Jerker Denrell and James G. March, 'Adaptation as Information Restriction: The Hot Stove Effect' in March (ed.), *Explorations in Organizations*, Stanford University Press, 2008, p. 139.
24 Alexander Cooley, *Logics of Hierarchy: The Organization of Empires, States, and Military Occupations*, Cornell University Press, 2005, p. 45.
25 March, *Pursuit*, pp. 124–6.
26 Stephen Biddle sees this set of core competencies as constituting the 'modern military system'. See Biddle, *Military Power: Explaining Victory and Defeat in Modern Battle*, Princeton, NJ: Princeton University Press, 2004.
27 Theo Farrell and Stuart Gordon, 'COIN Machine: the British Military in Afghanistan' in *Orbis* 53:4, 2009, pp. 667–9.
28 In fact, the commitment of the task force to southern Afghanistan was to breach British Defence Planning Assumptions; British defence chiefs were anticipating a drawdown of U.K. forces in Iraq which would release assets for deployment to Helmand. Testimony of General The Lord Walker (Chief of the Defence Staff, 2003–2006) to the Iraq Inquiry, 1 February 2010, pp. 55–7. Online. Available at: at www.iraqinquiry.org.uk/transcripts/oralevidence-bywitness.aspx
29 Sean Rayment, *Into the Killing Zone*, London: Constable, 2008, p. 33.
30 This was an extraordinary intelligence failure, especially as the U.S Ambassador in Kabul and the Commander of Combined Forces Command, Afghanistan, were both warning of a rising insurgency in 2006 and predicting that the Taliban would 'strike hard before NATO could become well established on the ground' in the south. Ronald E. Newmann, *The Other War*, Washington D.C.: Potomac Books, 2009, p. 52.
31 Patrick Bishop, *3 Para: Afghanistan, Summer 2006*, London: HarperPress, 2007, pp. 34–6; Rayment, *Into the Killing Zone*, pp. 37–40; Colonel Stuart Tootal, *Danger Close: Commanding 3 PARA in Afghanistan*, London: John Murray, 2009, pp. 23–4, p. 47.
32 James Fergusson, *A Million Bullets*, Bantham Press, 2008, pp. 164–5; Bishop, *3 Para*, pp. 49–50, 147–8, 178–9.
33 POR, UKTF/3005/H4, 19 October 2006.
34 POR, UKTF/H5, 11 April 2007, 2–0-4, paras 2 and 3.
35 POR, H5, 2–0-4, para. 14.
36 Ewen Southby-Tailyour, *Helmand, Afghanistan*, Ebury Press, 2008, pp. 77–8.
37 POR, H5, 2–0-3.
38 POR, H5, 2–0-3; 2–4-1; 2–4-2.
39 The Green Zone is the belt of fertile ground, between one and five miles wide, that runs the length of the Helmand River all the way from Kajaki in the north to Kanashin in the south. The population in Helmand is concentrated in this area.
40 POR H6, 19 October 2007, 5.
41 POR H6, pp. 2, 6.
42 Author interview with Brigadier John Lorimer, Permanent Joint Headquarters, Northwood, London, 29 June 2010.
43 POR H5, p. 3.
44 Author interview with Brigadier Lorimer.
45 Stephen Grey, *Operation Snakebite*, Viking, 2009, pp. 61–5.
46 Commander British Forces, Op HERRICK 7, 'Counterinsurgency in Helmand, Task Force Operational Design', TFH/COMD/DO7, 1 January 2008, p. 2
47 Annex A to COMD/DO 7, DTD, 1 January 2008, para. 8.

48 See note 39.
49 Author interview with Major Geoff Minton, 52 Brigade HQ, Redford Cavalry Barracks, Edinburgh, 29 June 2009.
50 Mackay, 'Making', p. 12.
51 Author's telephone interview with Brigadier Andrew Mackay, Commander 52 Brigade, 23 April 2008.
52 Q1: Have there been changes in the village population and why? Q2: What are the most important problems facing the village? Q3: Who do you believe can solve your problems? Q4: What should be done first?
53 Author interview with Brigadier Andrew Mackay, Commander 52 Brigade, 29 January 2009, DCDC Shrivenham.
54 Author interview with NKET officer for Operations Company in Laskhar Gar, 2 YORKS BG, 52 Brigade HQ, Redford Cavalry Barracks, Edinburgh, 29 June 2009.
55 Towards the end of its tour, 12 Brigade was providing 48 hours notice to locals in affected areas of planned major combat operations. Discussion with staff officer, 52 Brigade, Joint Services Command and Staff College, Shrivenham, 9 March 2010.
56 Author interview with SO3 J2 Plans, 52 Brigade HQ, Redford Cavalry Barracks, Edinburgh, 29 June 2009.
57 Tom Coghlan, 'Taliban Flee as Troops Retake Musa Qala' in *The Telegraph*, 10 December 2007. Online. Available at: www.telegraph.co.uk/news/uknews/1572108/Taliban-flee-as-troops-retake-Musa-Qala.html; Jason Burke and Richard Norton-Taylor, 'Allies Move into Town Held by Taliban' in the *Guardian*, 10 December 2007. Online. Available at: www.guardian.co.uk/world/2007/dec/11/afghanistan.military
58 Grey, *Operation Snakebite*, p. 280.
59 Post Operation Interview (POI) with Brigadier M. A. P. Carleton-Smith, Commander 16 AA Brigade, Op. HERRICK 8 (April–October 2008), 24 November 2008, Colchester, p. 2.
60 David Wilson and Gareth E Conway 'The Tactical Conflict Assessment Framework' in *RUSI Journal*, 154/1 (2009), pp. 10–15. TCAF has since been readopted by the British and is currently in use by 19 Brigade in Helmand. Telephone interview with Lieutenant Colonel Richard Wardlaw, CO 36 Engr. Regt., 9 February 2009.
61 Carleton-Smith, POI, p. 6.
62 Author interview with staff officer, 16 Brigade HQ, Merville Barracks, Colchester, 16 June 2009.
63 POR, TFH/H8/J7/13/11/03, 6 October 2008, Annex 2: Civil–Military Lessons.
64 The British Government's view in mid-2007 was that the Taliban did not present a 'strategic threat' to security in Afghanistan. House of Commons Defence Committee, *U.K. Operations in Afghanistan*, Thirteenth Report of Session 2006–2007, HC 408 (London: TSO, 18 July 2007), para. 35.
65 For example, in a survey of the lessons identified on the Land Lessons Database of this system from 2001–2008, I only found one that concerned the effectiveness of civil–military Comprehensive Approach. 'Stability operations and principles', 7th Armoured Brigade, TELIC 7, Land Lessons Database, Defence Lessons Identified Management System, accessed 29 April 2008 at U.K. Land Warfare Centre, Warminster.
66 Colin McInnes and John Stone, 'The British Army and Military Doctrine' in Michael Duffy *et al.* (eds), *Doctrine and Military Effectiveness*, Exeter: SPSG, 1997, pp. 14–25; Brian Holden-Reid, 'Warfighting Doctrine and the British Army' in Army Code 71565, *Army Doctrine Publication, Vol. 1 Operations*, HQDT/18/34/46, London: MoD, June 1994, 1A- 1–1A-15.

67 Colin McInnes, *Hot War, Cold War: The British Army's Way of Warfare, 1945–1995*, London: Brassey's, 1996, pp. 70–2.
68 POI with Brigadier A. D. Mackay, Commander 52 Infantry Brigade, Op HERRICK 7 (October 2007–April 2008), 3 July 2008, Redford Barracks, Edinburgh, p. 4.
69 Colonel Alex Alderson, 'The Validity of British Army Counterinsurgency Doctrine after the War in Iraq, 2003–2009', Defence Academy College of Management and Technology, Cranfield University, 2009.
70 Major G. E. Minton, PWRR, SO2 G5 Plans, HQ 52 Infantry Brigade, 'Operational Planning: A Reality', Powerpoint presentation notes, p. 2.
71 Farrell and Gordon, 'COIN Machine', p. 672.
72 Author interview with staff officer, Defence Intelligence, MoD, London, November 2008.
73 Sam Kiley, *Desperate Glory: At War in Helmand with Britain's 16 Air Assault Brigade*, London: Bloomsbury, 2009, pp. 33–4, 137.
74 Tom Coghlan, 'The Taliban in Helmand: An Oral History' in Antonio Giustozzi (ed.), *Decoding the New Taliban*, London: Hurst, 2009, p. 145.
75 16 Brigade reported the number of IEDs as 'doubling' each month of their tour (April–October 2008). Notes from Collective Debrief of 16 Air Assault Brigade, 3 December 2008, Merville Barracks, Essex. U.S. IED fatality rates in ISAF increased from 27 per cent in 2006 to 54 per cent in 2008, and 47 per cent up to May 2009. Jason H. Campbell and Jeremy Shapiro, *Afghanistan Index: Tracking Variable of Reconstruction and Security in Post 9/11 Afghanistan*, Brookings Institution, 19 May 2009, Figure 1.3, p. 6. Online. Available at: www.brookings.edu/foreign-policy/~/media/Files/Programs/FP/afghanistan%20index/index.pdf
76 Kiley, *Desperate Glory*, pp. 108–14.
77 Notes from Collective Debrief of 3 Commando Brigade, at RM Barracks, Stonehouse, Plymouth, 7 July 2009.
78 *Afghanistan Index*, Figure 1.15, p. 12.
79 ISAF Metrics Brief, 2007–2008, UNCLASS // REL USA ISAF NATO, slide 8.
80 Op. HERRICK 7, Op. Order, 27 OCT 2007, p. 1.
81 Author interview with Brigadier Andrew Mackay, Development, Concepts, and Doctrine Centre, Shrivenham, 29 January 2009.
82 Author interview with Brigadier Lorimer.
83 POI with Mackay, p. 2.
84 Mackay biography (September 2009), on file with author.
85 Andrew Mackay and Steve Tatham, *Behavioural Conflict: From General to Strategic Corporal: Complexity, Adaptation and Influence*, Shrivenham Paper No. 9, December 2009.
86 Richard Norton-Taylor 'Deadly and Maybe Decisive: Officers Hail Panther's Claw' in the *Guardian*, 27 July 2009. Online. Available at: www.guardian.co.uk/uk/2009/jul/27/panthers-claw-operation-afghanistan-taliban
87 Notes from Collective Debrief, 3 Commando Brigade.
88 POR, TFH/H10, 6 October 2009, 5.
89 Of course, successful adaptation by British forces in the field will not guarantee campaign success. Many other factors will shape the outcome of the campaign, including the effectiveness of ISAF in other provinces, and even more importantly, the growing capacities and effectiveness of Afghan public institutions and security forces.
90 The case of the U.S. military in Iraq from 2005–2007 also suggests the importance of studying bottom-up change by militaries at war. See James A. Russell, 'Improving in War: Counterinsurgency Operations in Anbar and Ninewa, Iraq, 2005–2007', Chapter 10 in this book; Philipp Rotmann *et al.*, 'Learning Under

Fire: Progress and Dissent in the U.S. Military' in *Survival* 51:4, 2009, pp. 31–48.

91 Author interview with SO1 Influence, Colonel Ed Flute, TFH HQ, Lashkar Gah, 3 October 2009.

92 Army Code 71876, British Army Field Manual, Vol. 1, Part 10, *Countering Insurgency*, London: MoD, October 2009, para. 1–3. This AFM also includes a whole chapter (Section 6) on Influence.

93 Lieutenant Mike Martin, 'The Importance of Cultural Understanding to the Military' in *British Army Review* 147, 2009, pp. 44–9.

94 Author interview with MoD official, London, 29 January 2010.

95 *The Significance of Culture to the Military*, Joint Doctrine Note 1/09, DCDC, January 2009.

96 John A. Nagl, *Learning to Eat Soup with a Knife: Counterinsurgency Lessons from Malaya and Vietnam*, Chicago IL: Chicago University Press, 2002.

97 Victoria L. Nolan, 'Command and Culture: Military leadership and the Evolution of the British Army's Approach to Small Wars', Ph.D. thesis, King's College London, 2009.

98 My thanks to Gautam Mukunda at MIT for suggesting this point.

99 John Kiszely, 'Learning About Counterinsurgency' in *RUSI Journal* 151:6, 2006.

10 Innovation in the crucible of war

Counterinsurgency operations in Anbar and Ninewa, Iraq, 2005–2007

James A. Russell

This chapter analyses operations by three battalions conducting counterinsurgency, or COIN, operations in Iraq in the period from July 2005 through March 2007: the 1st Battalion, 7th Marine Regiment (1–7) along the Iraq–Syrian border in the first half of 2006; the 1st Battalion 37th Armored Regiment (1–37) battalion operating in south–central Ramadi in the fall of 2006; and the 2nd Battalion, 1st Infantry (2–1), operating in eastern Mosul in 2005–2006. These operations occurred during a period of the war when the insurgency had reached its nadir, and insurgent groups exercised de facto political and military control over large areas of Iraq – particularly in Anbar Province. Many believed that the United States military faced the prospect of battlefield defeat and strategic disaster in Iraq during this period.

This chapter addresses two key questions: how did the units adapt to the growth of the insurgency and what were the sources of tactical adaptation and innovation in combating the enemy? The empirical evidence presented in these cases suggests that, contrary to popular perceptions, the units successfully innovated in war – a process largely executed organically within the units themselves. Innovation is defined here as the development of new organizational capacities not initially present when the units deployed into the theatre. The evidence presented in these cases suggests that the innovation process enabled these units to successfully transition from organizations structured and trained for conventional military operations to ones that developed an array of new organizational capacities for full spectrum combat operations. The units studied in this chapter developed these new capacities largely on their own initiative, drawing upon myriad sources of information and processes as they sought to improve their organizational performance in war. The process of innovation in these units was developed organically within the units, but, importantly, drew upon established training and doctrine to build their new organizational capacities on the battlefield. These case studies suggest a complex model of battlefield innovation involving both top-down and bottom-up processes as the units adapted and, as argued here, innovated while close with the enemy.

In the cases of the units studied here, the search for tactical solutions to the problems presented by the insurgency proceeded for the most part without interference and/or guidance on how to conduct counterinsurgency operations from higher headquarters at Multi-National Forces Iraq, or MNF-I – or any other military headquarters elements that might have dictated battlefield tactics. Moreover, there is no evidence that the political leadership in the Defense Department or other executive branch agencies sought to impose solutions at the tactical level – although MNF-I Commander General George Casey clearly faced political pressure from Secretary of Defense Donald Rumsfeld to avoid 'Americanizing' the war. Despite the pressure on Casey, however, no school solution materialized (as was the case in Vietnam) that allowed systemic biases at senior levels to impose themselves on commanders leading engaged forces. Ironically, the lack of a school solution can be explained partly by the lack of a doctrinally approved joint approach to fighting an insurgency before December 2006, when the Army and Marine Corps jointly released a draft of *FM 3–24 Counterinsurgency*. While the Army had released an interim doctrinal manual, *FMI 3–07.22, Counterinsurgency Operations*, in October 2004, the manual applied only to Army units and contained no guidance on structuring COIN operations at the 'joint' level. The interim manual also generally emphasized the kinetic side of COIN, focusing on destruction of the enemy's forces – as opposed to full spectrum operations.[1] These limitations mitigated the manual's impact in Iraq, since most military units faced a requirement to master full-spectrum operations and conduct tactical operations as joint task forces that incorporated units and organizational components from across the Department of Defense and other government agencies. Unconstrained by an established COIN doctrine, and impelled by desperation to find anything that worked, brigade commanders and the subordinate battalion commanders covered in this study were afforded wide flexibility in structuring their operations to fight the insurgency. As a result, commanders and their supporting organizations freely developed a series of new organizational capacities that helped improve their performance on the battlefield and reduce the effectiveness of insurgent operations directed at their units.

1st battalion, 7th marine regiment, Western Anbar Province

By the middle of 2005 military commanders described western Anbar as the 'epicenter of the country's deadly insurgency',[2] and reporters labelled the Euphrates valley the 'Ho Chi Minh Trail' of the war in which money, men, and material filtered in across the Iraq–Syrian border, down the valley, and all the way to Baghdad.[3] While Al Qaeda–Iraq (AQI), gradually seized control over many areas in western Iraq in 2004 and 2005, their presences in towns like Al Qaim were noticed by local residents. Over the period a tactical alliance emerged between AQI and Sunni insurgent nationalist groups bound together by their opposition to the occupation and fear of growing Iranian Shi'ite influence throughout the country. Importantly, however, many of the foreign AQI insurgents remained outside the area's indigenous tribal and social structure. The Iraq–Syrian border region had long been controlled by the Albu Mahal tribe, one of the main Sunni groups comprising the Dulaym tribal confederation in Anbar. While the Albu Mahals never enthusiastically supported Saddam, members of the tribe nonetheless gradually gathered under various insurgent nationalist groups opposing the occupation.

While initially supportive of AQI in late 2004, by the middle of 2005 evidence appeared suggesting that the Albu Mahals (as well as other tribal groups) had become uncomfortable with the relationship.[4] In a pattern that would be repeated elsewhere in Anbar in 2006 and 2007, and what would later become known as the tribal awakening, local tribes gradually came to object to AQI's presence. By the middle of 2005 the Albu Mahals and the Albu Nimr tribe had established their own militia – called the Hamza battalion – to actively resist AQI in the area along the Iraq–Syrian border. In mid-2005 press reports indicated that the Hamza battalion had assisted U.S. forces with intelligence tips during Operation Matador in May 2005.[5] In response AQI entered into an alliance with two rival tribes, the Karguli and Salmoni, over the summer, and finally drove the Albu Mahals from Al Qaim in September 2005.[6] In early September AQI had taken over Al Qaim and posted a sign that read: 'Welcome to the Islamic Republic of Qaim.' The Albu Mahals would later join with coalition forces in a unit called the 'Desert Protectors' in supporting the coalition offensives throughout the fall of 2005. They would be rewarded at the conclusion of the offensives and re-inserted into their positions of power and influence by coalition forces.

After the departure of 3rd Battalion, 6th Marines (3–6) in the spring of 2006, 1st Battalion, 7th Marine Regiment (1–7) commanded by Lieutenant Colonel Nick Marano, deployed in the fall of 2005 into the Al Qaim area to consolidate the 'hold' phase of the COIN campaign following the conclusion of Operation Steel Curtain in Western Iraq. The approaches of 3–6 and 1–7 built on each other as the process of adaptation gathered momentum through 2006. Upon deploying into western Anbar in the spring of 2006, Marano spread the battalion into the dozen odd company- and squad-sized outposts that had been established by 3–6 in the fall of 2005. Marano sought to continue the process, initiated by 3–6, of pushing the unit out into remote areas to consolidate the Marines' presence throughout the area.[7]

As 1–7 dispersed through western Anbar it built on the progress of 3–6 in generating local intelligence networks. By April of 2006 the locally oriented joint Iraqi Security Force' (ISF) Marine patrols generated nearly 80 per cent of the unit's intelligence.[8] The organic generation of intelligence initially overloaded 1–7's standard S-2 complement of between four and six officers. By end of its deployment, the 1–7 S-2 staff had grown to more than 30 analysts. The battalion's efforts to build organizational capacity to meet the demands for intelligence collection and analysis represented a critical component in its COIN campaign. The 1–7 intelligence effort represented a multifaceted program built on a wide array of technologies, new software and database programs, and perhaps, most importantly, a shared understanding throughout the unit of the importance of gathering data on a systematic basis to build situational awareness.

As a first step the unit organized a census and vehicle registration program for all the towns under its control.[9] Over the objections of Marano's headquarters the battalion established vehicle checkpoints at two locations in the area. Every vehicle had to be registered with the local police and a colour-coded sticker system identified the town from which the vehicle originated. The system allowed joint patrols to clearly identify unregistered vehicles and vehicles that were not from the immediate patrol areas. The vehicle registration system represented one component in an area-wide census conducted by 1–7 throughout early 2006. The unit divided each town into discrete named neighbourhoods; each street received a name, and each house received a number to allow 1–7 and its surrounding units to have a common frame of reference.

The census data proved critical to 1–7's operations throughout its deployment. Information gathered in the census, including the vehicle registration, was entered into to database called COPLINK, a law enforcement database and analysis software program used by the Phoenix Police Department that used artificial intelligence to simultaneously fuse information from multiple databases. Its appearance with 1–7 in Al Qaim was no accident. Marano's use of expertise from the American law enforcement community happened as a result of a series of programs administered in the Pentagon that sought to investigate the utility of building new COIN procedures that drew from the experiences of police departments around the United States. An organization called the Technical Support Working Group (TSWG) convened a series of workshops in the summer of 2005 to familiarize major metropolitan police departments with the Iraq COIN environment and to determine whether their law enforcement experiences could be of assistance.[10]

The workshops helped the Marine units to adapt law enforcement training and TTPs (tactics, techniques and procedures) suitable for the Iraq COIN environment. During 1–7's pre-deployment training the unit worked extensively with the Los Angeles and Phoenix police departments to help the unit develop a police-like set of tactics, techniques and procedures built around skills for observation, profiling, and questioning.[11] The training sought to make individual Marines comfortable with thinking about their jobs as similar to that of a policeman on the beat. Marano believed that a frame of reference based on gang warfare might prove useful in combating insurgent networks in Al Qaim. The battalion drew extensively upon the expertise of Ralph Morten, a Los Angeles police detective who spent several months advising 1–7 in Iraq. Morten, a 27-year veteran of the LAPD, was regarded as one of the nation's top experts on suicide bombings following several years spent training with the Israeli police.[12] Morten helped the unit secure $2.5 million from the Pentagon to fund a partnership with Lockheed–Martin that developed into an initiative called 'Project METRO', or Mobile Embedded Target and Reconnaissance Operation, which fused together the capabilities of the COPLINK database with new training based on tactical TTPs and a suite of surveillance equipment provided by Lockheed–Martin Corporation.[13]

'Project METRO' fused together a series of disparate capabilities and technologies that shaped 1–7's approach to fighting in the COIN environment in Al. The battalion staff inputted data gathered in the census and vehicle registration efforts into COPLINK and deployed sensors and cameras in areas identified through pattern analysis of attacks. All the information was queued to the battalion operations centre to provide an integrated common operational picture that helped support kinetic and non-kinetic operations.

The suite of sensors and data processing capabilities blended with the new TTPs developed for 1–7's COIN operations. The surveillance helped monitor insurgent activity in remote areas, and information collected in the program greatly assisted in link–nodal analysis of the insurgent networks. The unit installed overt and covert surveillance equipment throughout urban and other areas with high IED activity. The centrepiece of the system was a series of police surveillance cameras installed throughout the border city of Husaybah, which had a population of 120,000. The partnership with Lockheed–Martin brought additional surveillance equipment such as acoustic recorders that were left inside target houses suspected of supporting insurgent activity. These programs created a symbiotic effect: Marine and Iraqi foot patrols with enhanced skill sets tailored more towards law enforcement than traditional massed fire conventional operations; better intelligence collection and the enhanced situational awareness that resulted from the patrols; and a flexible battalion command element set up both to receive information and push information down to the lowest levels, quickly, in support of operations. The flattened organizational hierarchy created with the free flow of information up and down the command chain would be repeated elsewhere by other units discussed in this chapter.

Morten's deployment with 1–7 and the use of law enforcement technologies and TTPs in the COIN environment, were judged to be a great success. In its report to Lockheed–Martin Corporation on the utility of its suite of sensors, 1–7 noted that while not all the technologies worked well, the COPLINK database and supporting sensor suite 'vastly [reduced] the time necessary to create a target package from several hours to several minutes' and that the system was extremely useful in counter-IED operations.[14]

The law enforcement TTPs helped tailor 1–7's patrol techniques. One resulting new TTP assigned specific neighbourhoods to specific daily patrols. The unit reported that the new TTPs:

> allowed for them to develop the 'cop on the beat' mentality, knowing their terrain, and more importantly, knowing the people of the area. The level of familiarity gained with the area and its inhabitants allowed for easier detection of suspicious activity/ items/ personnel. More importantly, these tactics greatly facilitated the relationship with the local population, providing them with a much greater sense of security and willingness to provide information.[15]

Figure 10.1 shows data that tracks the slow but steady reduction in violence in and around Al Qaim from December 2005 to July of 2006. IEDs remained a persistent problem for 1–7 throughout its deployment. 1–7 focused extensively on the IED supply chain, a tactic which gradually reduced insurgent attacks during the deployment.

In March 2006, 1–7 stepped up recruiting local Iraqis for the ISF. Nearly 400 appeared on March 27, 2006, to join up. According to Marano: 'A lot of these guys were insurgents. It wasn't long ago we were shooting at them.'[16] The focus on building the ISF occurred simultaneously with the build-up of the local police force that had been started by 3–6 in the fall of 2005. Also building on the local leader engagement efforts mounted by 2–6, 1–7 successfully created a series of new police stations in and around the unit's outposts throughout the area. By the end of the spring the unit

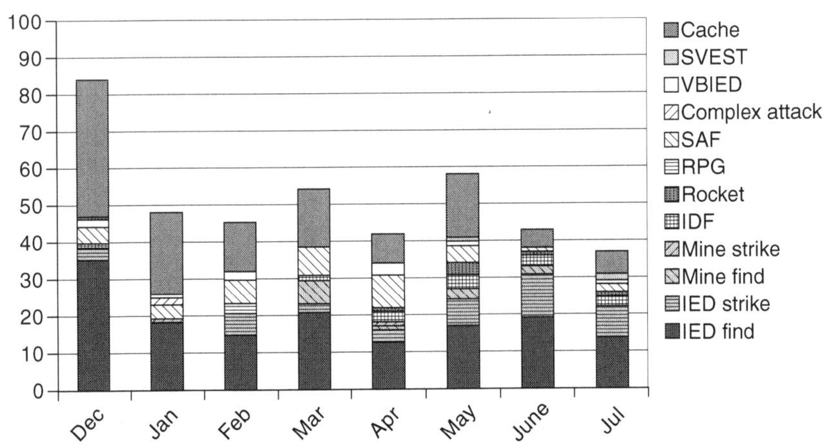

Figure 10.1 Incidents (source: PowerPoint presentation entitled 'Al Qaim PME', given by Lieutenant Colonel Nick Marano, then commanding officer of the 1st Battalion, 7th Marine Regiment, given at the Marine Corps Education Command, 21 February 2007).

had built a police force totaling 1,400 in the Al Qaim area. After receiving support from the local tribal leadership, 1–7 set up new Iraqi police stations in Husaybah, Ubaydi, Karabila, Sa'dah and Al Qaim. Marano stationed the police forces near U.S. outposts along with Iraqi security forces that steadily increased in number throughout 1–7's deployment. Concurrent with the focus on setting up the local police, developing the ISF, and conducting counterinsurgency operations, Marano's unit focused intensively on reconstruction and infrastructure in the towns throughout the area. The unit rebuilt schools, water treatment plants, health clinics, recreation centres, roads and even soccer fields. Two of the most important of these were the construction of two bridges over the Euphrates that had been destroyed during heavy fighting in 2005. The bridges were vital to restore local commerce in the Al Qaim region and western Iraq.

By the spring of 2006 the environment in western Iraq had improved dramatically. By this point in the campaign Colonel Blake Crowe, commander of all the Marines in western Anbar, called the area around Al-Qaim 'the model for where they want us to go'.[17]

1–37 in south–central Ramadi

1–37 served as part of the 1st Brigade Combat Team, 1st Armored Division (1/1) 'Ready First' combat team, or RFCT 1, based in Germany. During 1/1's deployment in Ramadi from July 2006–March 2007, the struggle with AQI and Iraqi nationalist insurgent groups reached its crescendo. The fight progressed week-by-week, block-by-block through the city's neighbourhoods as 1/1 moved systematically through the entire city applying the kinetic and non-kinetic tools at its disposal to retake the city. The 1/1 campaign in Ramadi, highlighted by the 'Awakening' process with Sunni tribes turning against AQI, will go down as a textbook example of successful COIN operations that its commanding officer, Army Colonel Sean MacFarland, later referred to as a 'three dimensional game of chess'.[18]

The brigade deployed into Iraq in January 2006, relieving the 3rd Armored Cavalry Regiment (ACR) in Tal Afar, Ninewa province. The 3rd ACR had conducted a much publicized and somewhat successful COIN campaign in the city, masterminded by Army Colonel Herbert Raymond McMaster.[19]

Upon arriving in Ramadi in July, 1/1 mounted aggressive local engagement initiatives and stepped up police recruiting efforts that proceeded simultaneously with sustained conventional military operations. The initial fight in Ramadi featured high-intensity kinetic operations. Of note, 1/1 deployed into Ramadi as an armoured regiment with equipment (M1A1 tanks and Bradley Fighting Vehicles) not traditionally thought of as relevant to successful COIN operations. The 1–37, however, found its armoured vehicles extremely useful in its battle with AQI. When the unit made contact with insurgents it used its Bradley vehicles to manoeuvre troops with precise, direct, fire support that could quickly eliminate insurgent forces in fixed locations. The vehicles also proved instrumental in securing the lines of communication and resupply between the combat outposts (COPs), established during its operations.[20]

After arriving in July 2006 1–37 and its task force elements assaulted south–central Ramadi – a part of the city that had been off limits to U.S. personnel. The campaign began with 1–37 isolating its sector by cutting off insurgent ingress and egress routes.[21] The unit began the campaign by constructing two initial COPs – Iron and Spear – which it used to expand its control over its sector. The insurgents initially responded to the establishment of the COPs with platoon-sized attacks on the outposts that 1–37 repulsed. Heavy insurgent losses in these attacks crippled their ability to launch future large-scale assaults.

Tedesco established three priorities for the unit. First, he wanted to seize the physical terrain in order to secure the population and disrupt insurgent operations. Establishing combat outposts provided 1–37 with the ability to establish a constant presence in the contested neighbourhoods and seize the physical terrain from the insurgents. Second, he planned to expand his unit's physical control over the terrain through clearing operations and aggressive area ambushes of insurgents. Gaining control over the terrain was supported through census operations moving through each block, building the unit's database detailing the neighbourhood resident population and their vehicles. Third, Tedesco sought to build ISF capacities to help consolidate control over the cleared areas.[22] Tedesco believed that the city's population represented the critical terrain for the battalion's operations. Operations of the unit were grounded in doctrinal processes, but the plan to retake the city reflected Tedesco's thinking on how the unit could adapt its kinetically oriented capabilities and training into Ramadi's COIN environment to meet the requirements of MacFarland's campaign plan. The battalion steadily expanded its control over its sector in south–central Ramadi from July 2006 through early 2007.

The unit cleared its neighbourhoods in successive chunks through the establishment of COPs, beginning with the establishment of COPs Iron, Spear and Falcon in June. Each of the COPS provided bases from which to start patrols and establish the continuous presence as envisioned by McFarland's campaign to wrest the neighbourhoods from insurgent control.[23]

Seizing the physical terrain became the means for 1–37 to address the critical component of the battle – securing each neighbourhood population. Over the course of the nine month campaign, 1–37 built six COPS and ten long-duration observation posts and secured the streets connecting these sites.[24] Over the course of its nine months in Ramadi the unit continued to refine and develop its TTPs on COP construction, building complex procedures that coherently synchronized necessary tasks throughout the battalion. By the end of its deployment, 1–37 could construct a COP in less than 24 hours. The COP construction procedures developed by 1–37 and other units in 1/1 were sent by the brigade to General Petraeus after he arrived in the spring of 2007. The procedures and materials used by 1/1 later became the basis for the 'COP in the Box' routines that were distributed throughout U.S. units in Iraq.[25] The unit conducted an estimated 3,200 combat patrols and mounted 275 company-level operations, killing an estimated 480 insurgents.[26]

While COP construction proved to be a critical component in 1–37's COIN campaign, the units developed a host of new procedures to establish local security over its ever-widening swathe of territory in the city. Simultaneous with COP construction, the unit concurrently developed and executed a census plan to build local situational awareness. Data derived from the census operations subsequently became critical to shaping many of the unit's daily operations.[27]

The battalion staff conceived the idea of a 'census loop' to link census activities with broader unit operations. Census data directly supported patrols, which in turn helped develop local human intelligence networks that provided more information for the census database. The census patrols became an invaluable instrument to help 1–37 peel back the complex layers of the neighbourhoods' social environment in south–central Ramadi. The idea for a census database started with one of the unit's company commanders, Captain Greg Pavlichko,[28] who commanded Company C deployed to the area of Ramadi around COP Spear. After establishing COP Spear, Pavlichko sent patrols to take pictures of the neighbourhood inhabitants and linked the pictures to PowerPoint files with overhead pictures of the houses on each block. The effective use of data gathering became apparent following an IED attack that killed an Abrams tank gunner, Sergeant Mark R. Vecchione, on July 18, 2006. In trying to track down the IED cell that executed the attack, the battalion intelligence section had a source that provided names of cell members. Pavlichko's database contained not only the pictures of the cell members by name, but also identified the exact location of their residences. 1–37 quickly detained 11 of the 12 members of the attack cell with minimum use of force.[29]

As the value of the initiative filtered up the command chain, the battalion quickly spread the census data gathering procedures throughout the battalion.[30] As a first step the unit refined the Ramadi's street and house numbering system, which it distributed to all unit members to create a common understanding of the area. The unit created questionnaires to collect information, it identified a Microsoft database to store and manage that information, and devised new tactics, techniques and procedures for the patrols that would conduct the census. The 'census patrols', as they were called, consisted of 15 to 30 soldiers. Each patrol covered eight to ten houses per patrol, spending between ten and 30 minutes in each home.

The 1–37 staff developed a standardized, bilingual, census questionnaire to be filled out either by the patrol leader's interpreter or the local residents. A completed questionnaire provided the patrol sector, building number, date visited, full name (including tribal affiliation), date of birth, occupation and location of work for each military-age male in the household. It also listed the number of women and children living in the house along with the religion of the household, whether the house was owned, rented or if the residents were squatters. The form included the serial number for the household AK-47, and whether or not the house had power and water available. The census questionnaire also collected the license plate number and description of any vehicle owned by the family. Upon completion the census patrol photographed all military-age males in the household with the subjects holding up their identification card and a placard with their name and house number.[31]

Tedesco and his staff realized that the census patrols represented far more than typical reconnaissance operations, which conventionally oriented Army training and doctrine viewed as a tool to prepare the battlefield for larger kinetic operations. In Ramadi 1–37 reversed this order, using larger scale cordon and searches as shaping operations for the census–reconnaissance patrols that would follow later.[32] The SOP for the unit typically focused on: (1) Planning and establishing the COP; (2) Ensuring route security so each outpost could be kept resupplied; (3) Clearing operations after the COP had been stood up to clear IEDs and find weapons caches; and (4) Census patrols to follow after the clearing operations to consolidate the position and gradually work its way into the human terrain of the area – the real target of MacFarland's campaign.

The unit came to grasp that the census data could contribute more to the long-term success of its operations than kinetically oriented cordon and search operations. Like the 1–7 Marines in western Anbar, the census patrols in Ramadi increased the face-to-face contact between the U.S. forces, the ISF and local communities, and began to generate organically driven intelligence on insurgent cells and weapons caches. The census also provided a structured way for local residents to file claims for any damage that might have occurred during cordon operations. Patrol leaders provided a claims card to the family and instructed them to deliver the claim to one of the task force's three Civil–Military Operations Centers for processing.[33]

As the patrols collected data two final tools helped to further exploit the information. Tedesco's staff developed a Microsoft Excel spreadsheet to compile the questionnaire data; one worksheet listed information about the house and a second worksheet compiled information specific to each military-age male. The database refined Pavlichko's initial product built with Microsoft PowerPoint. A hyperlinked presentation allowed one to click on a patrol sector, then a building, to show the pictures and names of all the military-age males, vehicles or any additional pictures of importance (a suspicious hole in the wall, the observation from the rooftop, etc.). Each company team was augmented by an intelligence analyst to help properly compile the census products. Each company then briefed progress in filling out their census databases at the weekly battalion-level targeting meeting to ensure compliance and completion.

In conjunction with the census patrols 1–37 developed a 'small acts of kindness' initiative. Realizing the frightening effect a nocturnal visit by 15 heavily armed and armoured foreign soldiers had on the residents of a home, each patrol carried an assortment of small toys, candy, and several two pound packages of sugar. Patrol leaders discovered that these gifts helped reduce the tension of census visits and helped generate exchanges that increased their understanding of local dynamics. Patrols found that the bags of sugar – a scarce commodity in Ramadi – proved particularly effective in generating positive feedback from the home's residents.

While the patrols started out as instruments to collect intelligence data to support counterinsurgent operations, they eventually became vehicles to also conduct civil–military operations, information operations and the development of human intelligence networks. In short the census patrols evolved into the task force's main instrument for fighting the counterinsurgency. By the end of its deployment 1–37 had completed a census of 80 per cent of the buildings in south–central Ramadi.

1–37 deployed into Ramadi as an armoured battalion – a legacy unit organized and equipped for conventionally oriented fire and maneouvre missions. The unit demonstrated significant adaptive flexibility and built what can only be described as sophisticated and systems-oriented COIN capacities in executing its part of MacFarland's campaign to retake Ramadi. While the Army is particularly noted for a rigid command hierarchy and a campaign-style approach to warfare, this unit clearly demonstrated its capacity for learning and searched for optimal solutions, accepted disparate sources of information, and constantly sought to build its understanding of the operational environment. As demonstrated in the building of the census program and supporting series of COIN-related procedures that followed, initiative bubbled its way to the top of the organization and then back down again in the form of routinized procedures. This organizational characteristic allowed 1–37 to develop capacities and structures producing outputs that met the requirements of the complex environment.

2–1 infantry in eastern Mosul, 2005–2006

Mosul's location near the Syrian border and its traditional role in regional smuggling and trade clearly represented one reason why the city became a favoured location for insurgents in 2005 and 2006. The terrain represented another attractive feature. Mosul's urban landscape and the topography around the city provided a rich environment for the insurgents.

The urban environment featured an extremely dense population and a warren of winding streets and ancient buildings. The area around Mosul contains many forests and groves as well as marshland on both sides of the Tigris. These features provided plenty of cover to hide training areas and insurgent compounds.

The COIN fight in Mosul proved a particularly difficult challenge to the wheeled Stryker units of the 172nd Stryker Brigade (as well as its predecessor, the 1/25 Stryker Brigade). Mosul's dense urban terrain lacked a well-established, modern road network, which meant that insurgents could quickly melt back into neighbourhoods criss-crossed by a series of winding alleyways too narrow for the Strykers to mount easy pursuit. The 2nd Battalion, 1st Infantry (2–1) commanded by Colonel Charles Webster, operated on the eastern side of Mosul. The battalion's 800–900 soldiers divided into two infantry companies, a cavalry troop, and 74 Stryker vehicles that conducted operations in an area hosting a population numbering 750,000–800,000 spread over five to six square miles.

Prior to arriving in Mosul, 2–1 actively participated in brigade-wide activities to build knowledge and understanding of COIN. The lack of established joint doctrine presented no hindrance to the unit as it prepared for its deployment. Using inputs from a wide variety of sources the battalion staff collectively identified goals and end states to achieve: a secure populace; established local political institutions; a contributing local government; neutralized insurgent capabilities; and, information flows from all sources.[34] In its pre-deployment training, the battalion organized its operations into four broad categories: security, governance, economics and information operations. To relate its operations to its end states, the battalion staff wrote its own campaign plan. While campaign plans are doctrinally the purview of corps- and division-level units in the Army, Webster saw value in an overarching plan to widely disseminate in the unit so that all members of the organization understood how their activities contributed to the campaign objectives. The plan drew upon Army doctrine, *FM 7.0 Training the Force,* which provided a methodology for building a campaign plan based on the mission-essential task list (METL).[35] These were core tasks that the unit judged it would need during the deployment. The METL formed the basis for the unit's pre-deployment training. The plan served as the battalion's template to guide its lethal and non-lethal targeting operations in attempting to meet the quarterly desired end states. The targeting cycle evolved over two week periods, in which the staff evaluated the non-lethal and lethal targets and then formally proposed the target sets to the battalion staff for mission planning and execution.[36]

Not surprisingly the battalion realized early in its deployment that a plan drawn up in Fairbanks, Alaska (the unit's peacetime home) would need adjustment in Mosul. The need for change emerged after the first three months on the ground as the battalion built situational awareness, gathered reporting data, and applied the targeting methodology. The battalion evolved through a three-phased analytical process during the first three months: (1) evaluate the enemy and his tactics, techniques and procedures, or TTPs; (2) evaluate the terrain – physical, human and economic/social makeup; and (3) evaluate the unit performance by examining the application of combat power through TTPs and the use of enablers such as Iraqi police and Iraqi Army commandos. The battalion's initial threat assessment of eastern Mosul came from intelligence assessments of the city's neighbourhoods compiled by the Brigade S-2 staff, which drew heavily on the experiences of the prior unit, the 1/25. Those assessments revealed neighbourhoods immediately east of the Tigris River free from direct insurgent control – but surrounded by a series of insurgent support zones influenced primarily by Ansar al-Sunna (AAS) and Al Qaeda in Iraq (AQI).

During October–November 2005 the battalion faced an upsurge in violence as the insurgents sought to derail the upcoming December 2005 legislative elections. In addition to targeting the Iraqi Army and Iraqi police forces, the insurgents actively targeted Stryker vehicles in order to film the attacks, using the videos to swing popular support in the city against the United States and the local government.

Attack trends and pattern analysis revealed that insurgents focused their IED attacks along the two major roads in eastern Mosul – one of which fell upon the seam of sectors patrolled by 2–1's A and B Companies. The seam had been exploited by the insurgents during successful IED attacks on four Strykers during October 2005. The battalion responded to these findings by adjusting its battlefield framework and tactical patterns. First, the battalion eliminated all traffic along the road that had been used by insurgents to mount attacks on the Strykers. Second, the battalion established a multi-layered information, surveillance and reconnaissance, or ISR, plan to detect the insurgent response to the road closure. The plan included overhead video surveillance by the brigade's remotely piloted vehicle, beefed up signals intelligence interceptions using the Raven airborne system, and aggressive efforts to improve humint (human intelligence) networks in the neighbourhoods surrounding the attack areas.[37] Last of all the unit eliminated the geographic boundaries between its units – meaning that its companies no longer had sole responsibility for a defined battle space. Instead, eastern Mosul became divided up into a series of non-contiguous 'L' designated areas.

The changes adopted by 2–1 began to pay immediate dividends with the roll-up of several insurgent networks that had successfully attacked the battalion during September and October. The battalion further leveraged this tactical evolution with organizational changes to its tactical humint teams, or THTs. The experiences of 2–1 in enhancing its ability to generate and analyse tactical intelligence mirrored those of many other units in Iraq over the period, which built more robust S-2 sections and generated better tactical intelligence to drive COIN operations. The 2–1 successfully reoriented its THT capacities to generate useful information, which, in turn, fed into a fused intelligence–operations cycle used by the unit for the remainder of its deployment in Mosul. By the end of its deployment, humint sourcing drove nearly 90 per cent of deliberate operations conducted by the battalion.[38]

Initially the unit drew upon humint networks that had been developed by the Central Intelligence Agency (known as 'other government agencies', or OGA) in the city. The CIA had a trained cadre of senior, trained personnel who had more latitude (i.e. money) and greater experience in supporting informant networks than did the brigade/battalion THT personnel. The all-source information from established humint networks was fed into the battalion's databases to build its understanding of the insurgent networks.[39]

In addition the unit initiated changes in its internal organizational structure to generate and handle intelligence. First, the battalion pushed THT personnel directly down to its manoeuvre elements to participate in patrols and directed raids. The battalion soon discovered that on-site questioning produced the most useful information, which could then be passed in real time to the battalion S-2 for correlation and analysis. Detainees were often at their most vulnerable psychologically immediately after being picked up in cordon and knock operations, or after being pursued following engagement in actual operations. The quality of the information generated from interrogations further improved as a result of new field questioning and interrogation strategies developed by the battalion with OGA inputs. After pushing the THTs down into his manoeuvre companies, Webster placed the battalion executive officer, Lieutenant Colonel Richard Greene, in charge of detainee questioning, in effect bridging what had been a gap and lack of capacity between the operations (S-3) and intelligence sections (S-2) on the battalion staff in the conduct of detainee questioning. Approximately 500 detainees cycled through the 2–1 questioning process over the course of ten months. The battalion started seeing results to the increased role of OGA-sourced networks and the new detainee interrogation supervision system. In mid-September it had generated little information to populate its link–nodal analysis of the insurgent cell structure. But by October the gaps were being filled – a process that gathered momentum over the next several months.

Information about insurgent cell structures steadily grew in richness and detail during the fall of 2005, enabling the battalion to develop its intelligence–operations fusions cycle, and to integrate non-kinetic effects such as information operations and civil–military operations (CMO) into the targeting process. Information gleaned from interrogations was greatly aided by the participation of OGA, Special Forces and the ISF personnel during the fall of 2005. Detainees were well aware that the United States could only hold them for two weeks before being transferred into the Iraqi

National Government's detention process, after which they would often be released within weeks. The detainees most feared the prospect of being transferred to Iraqi police custody. When the battalion introduced ISF personnel into the detainee interrogation process, detainees invariably became much more cooperative in the interrogations.[40] All these changes led to populated 2–1 databases that fed a growing link-analysis chart of the insurgent network in eastern Mosul.

The changed TTPs, the ISR collection plan, and the focus on generating better humint led to concrete results in late October and November as 2–1 began to disrupt insurgent cells. In late October the battalion successfully targeted insurgent cells that had mounted attacks using IEDs, mortars and snipers. Using information from the ISR collection plan and humint developed information, the battalion mounted a simultaneous operation by two Stryker companies and a cavalry troop in the Al Sharkya neighbourhood of Mosul.

Over a 36 hour period a fused intelligence collection and operations cycle had significantly disrupted several entrenched insurgent cells. The operation had begun with deliberate and detailed target exploitation of detainees that revealed cellular linkages and enemy TTPs.

The development of the battalion's fused intelligence–operations cycle facilitated the integration of deliberate lethal and non-lethal planning in an attempt to 'mass' effects in particularly troublesome neighbourhoods. The objective of massing effects represented the attempt to operationalize the 'ink spot' strategy used by various military commanders in Iraq during the period to create zones of stability free from insurgent influence and control – seeking to gradually expand those areas over time.

The assessments of the first several months drove the battalion into an expanded data collection and analysis operation to try and improve the unit's situational awareness on a neighbourhood-by-neighbourhood basis. Using a software tool called the situational template, or SITEMP, the battalion began delving beneath the data populating its databases coming from SIGACTS that reported significant actions and comprised of data flowing from enemy attacks and the unit's response. The unit populated a Microsoft database of each neighbourhood focusing on: key political and religious leaders, content being delivered in the mosques, location and capabilities of local medical facilities, public works facilities, mass transit services, information on local business, and other details. While the databases did not constitute house-by-house census detail – a process used elsewhere in Iraq by U.S. units – the databases provided a product that could be passed along to the next unit. The neighborhood SITEMPs, as they were called, helped the battalion build situational awareness of the complex dynamics driving and/ or supporting the insurgent violence.

The unit's ISF partnership activities at brigade and battalion staff levels became operationalized early in 2–1's deployment, with Iraqi Army Units initially only observing the battalion's targeting process. In late 2005 that observation gradually evolved into active participation in the targeting process – which greatly improved the partnership (particularly in the 4th IA Brigade).

The battalion divided up its activities into its 'day' and 'night' jobs. Night jobs were usually directed raids conducted by battalion personnel, whereas their day jobs included training the Iraqis and working on creating other battlefield effects. Preparing the Iraqi Army to conduct joint operations fell under the rubric of the battalion's day job.[41] To facilitate the interaction with the Iraqis and build tactical proficiency, 2–1 formed tactical combat advisor teams, or TCATs, to operationalize their objective of building ISF capabilities. The TCATs, commanded by Captain Jason Glemser and Captain Rusty Topf, subsequently established a small unit training program for two Iraqi Army battalions. The TCAT mission was to 'train and advise the Iraqi infantry companies and below in order to facilitate their capabilities to conduct unilateral missions and the handover of combat missions from U.S. forces to the Iraqi Army'.[42] Each team consisted of five men – a sergeant first class team leader, three junior noncommissioned officers and a soldier. The TCATs designed and supervised a four week training cycle focused on basic combat skills for an Iraqi company, with one platoon a week in training.[43]

The battalion succeeded in reducing the effectiveness of insurgent attacks, which continued throughout the deployment at relatively constant levels. As indicated in Figure 10.2, total attacks mounted in eastern Mosul

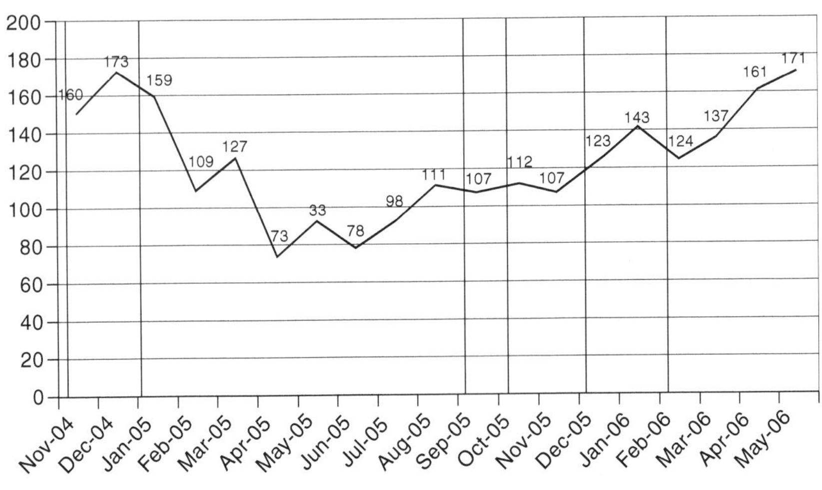

Figure 10.2 Attack Trends in eastern Mosul, November 2004–May 2006 (source: PowerPoint presentation entitled '3-2 Infantry Transition Brief', given by Lieutenant Colonel Charles Webster, then commander of the 2nd Battalion, 1st Infantry to the incoming unit the 3rd Battalion, 2nd Infantry, in June 2006).

actually increased slightly during the period of 2–1's deployment. Behind the numbers, however, had been a shift in insurgent tactics during the spring of 2006 and a decreasing lethality over the period. As the ISF steadily gained in competence during the spring of 2006, insurgent attacks increasingly focused on Iraqi Army and police personnel. The ISF became subjected to a concerted insurgent fear and intimidation campaign in the spring of 2006 that mirrored tactics used elsewhere in Iraq as the ISF slowly built its capacities. The growing integration of ISF into the battalion targeting cycle and improved operational capabilities led to a steady roll-up of insurgent cells. The continuous disruption of these cells meant less skilled bomb makers were established and the IEDs and SVBIEDs became less effective.[44]

The battalion's COIN campaign demonstrated a systematic attempt to apply the concepts of effects-based operations in a difficult environment. The battalion built a campaign plan from the ground up to apply a variety of organizational capacities to achieve its end states of a neutralized insurgency and stable local security, an ISF capable of independent operations and a government capable supporting the population's basic needs. It is clear that the battalion innovated in the field in its attempt to deliver effects to create that end state. The battalion had its greatest success in direct action against insurgent cells. The building of the unit's operations–intelligence fusion cycle created an extremely agile, flexible, organization that drove authority down the organizational hierarchy. The brigade's pre-deployment emphasis on junior leader development successfully created a middle management that accepted and exercised authority and initiative and flourished in the field. It is also clear that that in 2–1's case the unit worked hard to use non-lethal effects through information operations and local leader engagement. By the definition used in this chapter, the battalion demonstrated significant innovative capacity in the field in the absence of overarching joint doctrine.

Conclusion

The case studies in this chapter chronicle an iterative process of organically generated tactical adaptation and innovation that unfolded over time in a distinctive progression. The process began in what could be described as tactical, ad hoc adaptation in which individual leaders reacted to local circumstance by cycling through different ways of employing their units and equipment on the battlefield. Some of these adaptations succeeded and others failed. As leaders identified successful adaptations the process gathered momentum and new organizational standard operating procedures emerged that became more widely adopted in the units. Organizational innovation then manifested itself through the emergence of new standard operating procedures that led to fundamental changes to the ways in which the units in this study fought the insurgents over the course of their deployments. As these innovations produced success on the battlefield they eventually fed into more formalized military doctrine that followed later.[45] While execution of the innovation process happened in the field, it is clear the process involved many actors throughout the military chain of command. Individuals, units, and headquarters elements stretching from Al Anbar to Baghdad, Fort Leavenworth, the Pentagon, and beyond searched for solutions to the problems being encountered on the battlefields of Iraq.

The cases detail a process of wartime innovation that manifested itself as a series of organic, bottom-up procedures developed within the battalions and brigades fighting the insurgents. While the innovation process developed organically, that process drew upon information and enabling processes nested in a variety of sources outside the units themselves. Innovation occurred within the units through the fusion of the information and enabling processes that ultimately produced new organizational outputs. Basic capstone doctrinal grounding in military operations proved to be a fundamental building block for the innovation process. Previous experiences in Iraq and Afghanistan helped shape the innovation. Non-institutionally blessed sources of information on COIN theory and history were consciously drawn upon by units as they intellectually reoriented themselves towards full spectrum operations.

A variety of digital domain platforms helped units freely pass information and lessons learned back and forth, which helped the process of adaptation and innovation. The innovation produced new organizational capacities that shaped successful military operations across the spectrum of kinetic and non-kinetic operations to reduce insurgent-generated violence. Wartime innovation in the cases studied here flowed from agile, flexible, decentralized organizations that featured flattened and informal hierarchical structures. Throughout their deployments each of the units covered in the case studies demonstrated significant learning capacities that proved central to the innovation process. The cases present a picture of military organizations acting in ways that are contrary to the popularly accepted view that they function as bureaucratically inclined, hierarchically structured organizations slow to respond to changes to their external environment. In the cases studied here the exigencies of wartime produced much different organizational behaviour.

While the execution of the battlefield innovation process happened within the units in this study, it would be a mistake to strictly characterize that process either as 'bottom-up' or 'top-down'. The processes of wartime innovation in the units studied in this chapter did not result solely from factors within the unit, nor was the process solely prompted by a top-down process featuring the articulation of a new military doctrine, whether created by forceful civilian intervention from above or by dynamic senior military leadership at the headquarters level. The innovation process exhibited by the units in this chapter was dialectical in nature and drew upon a complex series of forces, both from within and outside the units that fused together in ways that produced organically generated change – change that eventually 'pulled' tactical practice, institutional innovation and (finally) authoritative doctrinal pronouncements along behind it.

Established doctrinal processes as described in foundational capstone doctrine like the Army's *FM 3.0 Operations* provided a framework and common understanding for the units as they responded to the insurgency. The Marine units also drew upon their established doctrine to help provide a framework to structure their COIN operations. All of the units in this study exhibited firm grounding in doctrinally bounded processes and applied those processes to the problem of COIN in Iraq. Instead of being hampered by rigid bureaucratic organizations the wartime experiences of these units showed that networked, informal, and cross-functional organizations sprang up over the course of military operations, which fused disparate organizational elements, both military and civilian, into a synergistic whole applied in their COIN operations.[46]

Evidence from these cases suggests that certain American ground units (both Army and Marine Corps) in Iraq evolved into flexible, adaptive organizations taking advantage of twenty-first century human and technological capacities. The units in this study proved to be technologically advanced, complex, organizations with a highly educated and trained work force that embraced environmental complexity and searched for optimal solutions to operational problems. The organizations covered in this study did not take the path of least resistance in their search for solutions to the tactical problems posed by the insurgency in Iraq.

The units discussed above built new innovative core competencies within their organizations, drawing upon such factors as: (1) digitally based communications and data systems that seamlessly passed information on a continuous basis between units preparing for their deployments; (2) imaginative battlefield leadership that delegated authority, welcomed the concept of distributed operations, encouraged the free flow of information throughout the organizational hierarchy, and freely changed its organizational structures in the field in order to apply its capacities across the spectrum of kinetic and non-kinetic operations; (3) the use of advanced technologies and analytical methods to combat insurgent networks; and, (4) a continuous education process in which units kept seeking information and expertise from many different sources to aid in their tactical decision-making processes.

The emergence of American tactical COIN competence adds an interesting twist to one of the popular narratives of the period. Prior to the Iraq War the Bush Administration initiated a process called 'transformation' to reform the Defense Department's sprawling military and civilian bureaucracies. This process overwhelmingly featured top-down direction from the Defense Department's civilian leadership to executing organizational elements. As is well known, Secretary of Defense Donald Rumsfeld sought to make the Iraq invasion an advertisement for a new American way of war featuring precision-guided munitions, speed of movement, and effects-based operations, based on the IT-RMA.[47] Rumsfeld's new way of war, however, had little to do with counterinsurgency and activities he derisively referred to as 'nation-building'.

The irony of the evidence presented in this chapter, however, is that the certain units indeed 'transformed' themselves during the Iraq War though not in the ways envisioned by Rumsfeld. Prior to December 2006 and the promulgation of the new doctrine the most important part of the 'transformation' process occurred not in the invasion but in the counter-insurgency campaign that followed, and not through top-down direction but primarily through ground-up, organic processes in which tactical units eventually embraced and mastered the very 'nation building' skills that Rumsfeld sought to avoid.

Notes

1 See David Ucko's discussion of the interim manual and other Defense Department initiatives to reorient the military departments towards irregular warfare and COIN: *The New Counterinsurgency Era*, Washington: Georgetown University Press, 2009, pp. 65–80.
2 Mark Mazzetti and Solomon Moore, 'Insurgents Flourish in Iraq's Wild West' in *Los Angeles Times*, May 25, 2005.
3 John F. Burns, 'Iraq's Ho Chi Minh Trail' in *New York Times*, June 5, 2005.
4 See, for example, Sabrina Tavernise and Dexter Filkins, 'Local Insurgents Tell of Clashes with Al Qaeda's Forces in Iraq' in *New York Times*, January 11, 2006. Also see details in Steve Negus, 'Border Region Offers Glimmer of Hope for Post Insurgency Peace' in *Financial Times*, May 6, 2006; Mazzetti and Moore, 'Insurgents Flourish in Iraq's Wild West'.
5 Hannah Allam and Mohammed al Dulaimy, 'Iraqis Lament Call for Help' in *Philadelphia Inquirer*, May 17, 2005.
6 Ahmed Hashim, 'Iraq's Civil War' in *Current History*, January 2007, pp. 2–10; Carter Malkasian, 'Did the Coalition Need More Forces in Iraq? Evidence from Al Anbar' in *Joint Forces Quarterly* (3rd Quarter 2007), p. 123.
7 Ibid.
8 John Koopman, 'Marines Helping to Line Up Sunnis for Iraq's Army'in *San Francisco Chronicle*, March 27, 2006.
9 Author interview with Colonel Nick Marano April 17, 2008.
10 Results of one workshop are summarized in an unpublished White Paper titled, 'Confidence of the Community: Law Enforcement Support to Counterinsurgency', dated June 27, 2005.
11 Author interview with Colonel Nick Marano.
12 H.G. Reza, 'Arming Marines With Know-How For Staying Alive' in *Los Angeles Times*, October 24, 2005.
13 Matt Hilburn, 'Policing the Insurgents' in *Seapower Magazine*, March 2006.
14 *Memorandum*, from 1/7 Marines Intelligence Section to Lockheed–Martin, October 2, 2006.
15 *Memorandum*, from Commanding Officer 1/7, Baker Company (Captain C. A. Wolfenbarger, USMC) to Commanding General 1st Marine Expeditionary Force, *Subject: OIF 05–07 Lessons Learned in Regards to Training Provided by Mr. Ralph Morten*, LAPD, October 4, 2006.

16 John Koopman, 'Marines Helping to Line Up Sunnis for Iraq's Army' in *San Francisco Chronicle*, March 27, 2006.
17 Charles Crain, 'Marines on the Beat in Iraq' in *Asia Times Online*, June 7, 2006.
18 Author interview with MacFarland, October 15, 2007.
19 As detailed by Major Neil Smith, 'Retaking Sa'ad: Successful Counterinsurgency in Tal Afar' in *Armor Magazine*, July/August 2007, pp. 26–35. Also see David R. McCone *et al.*, 'The 3rd ACR in Tal'Afar: Challenges and Adaptations' in *Strategic Studies Institute*, Carlisle PA: U.S. Army War College, January 8, 2008; Lawrence Kaplan, 'Centripetal Force: The Case for Staying in Iraq', *The New Republic*, March 6, 2006, p. 19. For other details on the 3rd ACR operations, see Tal Afar.
20 Author interview with Colonel V. J. Tedesco, June 8, 2009.
21 Dexter Filkins, 'U.S. and Iraq Retake Ramadi One Neighborhood at a Time' in *New York Times*, June 27, 2006; Monte Morin, 'Taking Up Residence in Insurgent Havens' in *Stars and Stripes*, August 13, 2006; Monte Morin, 'Unexpected Neighbors Bring Hope in Ramadi' in *Stars and Stripes*, August 24, 2006.
22 TF 1/37, '"Bandits" Coin Operations in Ramadi', PowerPoint Brief (undated, kept by the author).
23 Fighting around COP Falcon in early August 2006 is detailed in Monte Morin, '1st AD Units Hit Insurgents Hard in Largest Battle of New Campaign' in *Stars and Stripes*, August 3, 2006.
24 A photo essay on COP construction by the 40th Engineer Battalion in Ramadi is detailed by Monte Morin, 'DIY Base Construction in Ramadi' in *Stars and Stripes*, August 21, 2006.
25 Author interview with MacFarland, October 15, 2007.
26 Ibid.
27 Ibid.
28 As recalled by Colonel Tedesco in author correspondence, and confirmed by Captain Greg Pavlichko, USA, Commanding Officer of C Company, 1/37 in author interview May 13, 2008.
29 Author interview with Captain Greg Pavlichko May 13, 2008.
30 Ibid.
31 Information on 1–37's census activities was derived from a variety of different sources: (1) Author interview with Colonel Tedesco; (2) Powerpoint briefing titled *TF 1/37 COIN Operations in Ramadi*; (3) unpublished paper by Captain Dave Black, Captain Jon-Paul Hart and Lieutenant Colonel V.J. Tedesco III, 'Sun-Tzu and BeanieBabies: Census Operations in Urban Counterinsurgency'. Paper cited with permission of Colonel Tedesco; (4) author interview with Captain Greg Pavlichko, Commanding Officer, C Company, 1/37, May 13, 2008.
32 TF 1/37 cordon and search operations are chronicled in Monte Morin's photo essay, 'Photo Gallery: Cordon and Search Operation in Ramadi' in *Stars and Stripes*, August 3, 2006.
33 Ibid.
34 Author interview with Colonel Charles Webster, March 7, 2009. Ideas further fleshed out in Lieutenant Colonel Wayne Brewster, 1–25 S3, and Lieutenant Colonel Charles Webster, TF 2 Senior JRTC Operations Group, 'Task Force Adaptive Planning: Everything is Connected', *A Battalion Task Force in COIN*, CALL Newsletter No. 08–25 (July 2008), Center for Army Lessons Learned, Fort Leavenworth, Kansas, 4.
35 *Field Manual 7–0, Training the Force*, Department of Army, Washington, D.C. (October 22, 2002)
36 Author interview with Colonel Charles Webster, March 7, 2009.
37 Details of the Raven system components available online at: www.telemus.com/datasheets/s-eagle.pdf and www.telemus.com/datasheets/raven.pdf.

38 Author interview with Colonel Charles Webster, March 7, 2009. Also see LTC Richard G. Green, Jr. TF XO, and CPT Mark N. Awad, TF 2–1, 'Optimizing Intelligence Collection and Analysis: The Key to Battalion-level Intelligence Operations in Counterinsurgency Warfare', *A Battalion Task Force in COIN*, CALL Newsletter no. 08–25, pp. 17–32.

39 Author interview with Colonel Charles Webster, March 7, 2009.

40 Green and Awad 'Optimizing Intelligence', pp. 17–32; Author interview with Colonel Charles Webster, March 7, 2009.

41 Author interview with Colonel Charles Webster, March 7, 2009.

42 Major Jason Glemser, 'Task Force Partnership', A Battalion Task Force in COIN, CALL Newsletter no. 08–25, July 2008.

43 Ibid; Author interview with Colonel Charles Webster, March 7, 2009

44 Author interview with Colonel Charles Webster, March 7, 2009.

45 Distinctions between these processes are drawn from a combination of works. See Michael D. Doubler, *Closing with the Enemy: How GIs Fought the War in Europe, 1944–1945*, Lawrence, Kansas: University of Kansas Press, 1994, p. 8; Stephen Peter Rosen, *Winning the Next War: Innovation and the Modern Military*, Ithaca: Cornell University Press, 1991, pp. 1–53; Barry Posen, *Sources of Military Doctrine: France Britain, and Germany Between the Wars*, Ithaca: Cornell University Press, 1984. pp. 34–80.

46 Thomas Ricks and Karen De Young, 'Al Qaeda Reported Crippled' in *Washington Post*, October 15, 2007, A1.

47 As chronicled by Bob Woodward, *Plan of Attack*, New York: Simon & Schuster, 2004.

11 Blitzkrieg, the RMA and defense intellectuals

Rolf Hobson

During the 1920s the Reichswehr under Colonel General Hans von Seeckt drew lessons from defeat in World War One that were very different from those of the victors, Britain and France. It sought to develop operational theories that would allow it to escape from the attritional stalement of trench warfare. A new force structure, combined with new tactics developed towards the end of the war placed a premium on flexibility and initiative. Mobility was emphasized as an alternative to firepower. Mechanization and motorization would restore the war of movement, which would once again make it possible to seek victory in decisive encounters. In spite of the restrictions imposed on Germany by the Versailles Treaty, the unofficial General Staff encouraged doctrinal innovation, tested its theories in manoeuvres and developed the required new weaponry through its secret cooperation with the Red Army.

The Reichswehr recognized early the potential of the tank and the importance of the new armour theories developed in Britain by Liddell Hart and J.F.C. Fuller. Heinz Guderian introduced them in the German Army. In marked contrast to the hide-bound conservatism of the French and British hierarchies, its open-minded military culture put a premium on intellectual suppleness, which fostered innovation. In the course of the 1930s these tendencies fused in the revolutionary operational doctrine of Blitzkrieg, which in essence entailed the use of independent armoured divisions, supported by air power, to break through the enemy's lines and force a rapid decision. Britain and France failed to understand the importance of these developments; their armies were stuck in a defensive 'Maginot mindset' that could conceive of no other role for tanks or planes other than for infantry support. The superiority of German doctrine was demonstrated in May and June 1940. The Netherlands and Belgium were occupied in a combination of airborne and armoured assaults. The world's most powerful army, the French, was utterly outmanoeuvred and defeated in a six-week campaign. France sought an armistice and Britain was expelled from the Continent. In one fell swoop Germany had come close to establishing a continental hegemony.

The picture drawn in the preceding paragraphs is not that of present day military historiography. It can be found, with little variation, in the vast outpouring of popular military history over the past six decades, but this will not be treated in the following. The purpose of this chapter is to identify that particular segment of strategic theorizing in which defense intellectuals draw lessons from past conflicts and apply them to what they believe to be the shape of future war.[1] It seeks in particular to identify some of the conceptual roots of the Revolution in Military Affairs (RMA), the leading theoretical concept among American defense intellectuals in the 1990s and the early years of the present century.

There can be little doubt that the description of Blitzkrieg given above represents the paradigmatic RMA. It is always cited as an historical precedent in the RMA literature.[2] More specifically, the elements which came together to produce the German revolution are identified as the integration of technological change through organizational adaptability and doctrinal innovation. The fusion of these three elements resulted in a vast increase in military effectiveness, and the proof of the pudding was provided by the battlefield victories of 1940. This process can and should be replicated, says the RMA literature, in an American revolution to provide the U.S. military with a superiority that no twenty-first century contender will be able to match.

The RMA interpretation of Blitzkrieg can draw on a large number of historical studies for support.[3] But there is also a more recent and growing body of research that questions many of the elements which together make up its image of Blitzkrieg.[4] Defence intellectuals have ignored the historical revisionism that does not fit their preconceived theories. Indeed not many of the canonical works on RMAs, which refer to past models, can be said to meet acceptable standards of historical scholarship, in the very restricted sense that they do not take account of all the relevant scholarly literature. Neither do they present reasoned arguments in favour of their understanding of military revolutions, or against alternative interpretations.

This does not necessarily disqualify the theories they advance. If strategy is a separate discipline that draws on historical illustrations and social science models, but asks different questions, it will have a methodology distinct from that of history or political science. Thinking about how to win future wars is not the same approach as trying to find out how they were won in the past, and requires different methods. But as strategic theories move further away from the historically verifiable, they move closer to the methodological trap into which defense intellectuals are prone to fall. That is, the tendency to project into the past the preferred doctrines of the present, and thereafter to read them out of history as lessons applicable to future warfare. The famous *Strategiestreit* between the civilian military historian Hans Delbrück and the Historical Section of the German General Staff centred on this problem.[5] The officer-historians of the latter believed that Fredrick the Great had conducted his wars in accordance with the perennial formula for victory: envelopment and the destruction of the enemy's force in decisive battle. Delbrück demonstrated that in practice Frederick deviated from this principle so often as to make it unlikely that he actually adhered to it.

In the following, the Blitzkrieg of RMA literature will not be compared with a 'correct' historical interpretation.[6] But some recent theories about the Wehrmacht's way of war and the fall of France will be presented. They can be mutually exclusive, and all they have in common is that they are drawn from the work of authorities in the field that are not discussed in the RMA literature. A selection among them will be combined into an alternative interpretative permutation very different to that of the RMA Blitzkrieg. The point here is not to win an historical argument, because one side – the defense intellectuals – is not engaged in one. It is rather to use an alternative permutation to uncover some of the unspoken assumptions that inform the war image of U.S. defense intellectuals. These deserve further discussion within the think-tank community, at least if it aspires to be a self-reflective and self-corrective discipline.

During the last 15 years something approaching a scholarly consensus has appeared, to the effect that the Wehrmacht did not have an operational doctrine in 1940, whether it was called Blitzkrieg or anything else. More precisely there were no official guidelines for the independent, strategic use of the Panzer arm.[7] Beyond this affirmation consensus starts to unravel, but there seems to be broad agreement on the following: the Wehrmacht's theory and practice in 1940 grafted the new technology of tanks, planes and radio communications onto traditional, nineteenth century operational ideals, specifically the *Umfassungsschlacht* that had found expression in the Schlieffen Plan.[8] The war of movement was espoused, not as an alternative to firepower but as the preferred option of the inferior power.[9] British armour theory probably influenced or speeded up the integration of new technology into previously existing operational ideals, but to what degree is the subject of discussion.[10] Without even taking into account parallel developments in other armies it is possible within the parameters of current research to state, in the words of Gil-li Vardi's recent dissertation, that: 'Ample evidence [...] suggests that the German army's activities in the 1930s did not fit any of the commonly described characteristics of an RMA.'[11]

Blitzkrieg was originally a descriptive term which entered currency when foreign journalists and military observers used it to explain Germany's staggering successes in 1939 and 1940.[12] After the unexpected collapse of France in the course of no more than six weeks, however, Hitler and the German military leadership believed that they had indeed invented a new kind of war. As Hitler put it to the head of the *Oberkommando der Wehrmacht* (OKW): 'Now we have shown what we are capable of. Believe me, Keitel, a campaign against Russia would be mere child's play [*ein Sandkastenspiel*] in comparison.'[13] Operation Barbarossa was consciously planned as a three-month Blitzkrieg. With these reservations, historians can justifiably use the term to encapsulate German operational thought, at least during the campaigns of 1941. And in a more general sense it can help to capture the dynamic of Nazi warfare as an attempt to break the bounds of Germany's constrained resources through conquest and plunder.

Historians may agree on the lack of a German operational doctrine prior to victory over France, but there is much less agreement over the causes of that country's defeat. The RMA literature generally takes it as given that the fall of France 'proved' the superiority of German doctrine over French doctrine. Alternative explanations are however available, and one could start by asking whether they might not qualify the importance assigned to doctrine. One could follow up by asking whether a comparison of doctrine that takes for granted the isolation of the operational level does not obscure more than it enlightens.

Doctrine vs. doctrine

It is now generally accepted that neither side in 1940 possessed a decisive superiority in either men or machines.[14] Yet although the Wehrmacht defeated the French army (and the BEF), it does not necessarily follow that this represented a victory of doctrine over doctrine. Manstein's *Sichelschnitt* ('the cut of the sickle') took huge risks, and few would deny that the French High Command made some colossal mistakes. As Denis Showalter puts it: 'Even Blitzkrieg depended on the existence of "obliging enemies" who would not only simply make mistakes, but who would make good mistakes.'[15] The two most obvious ones were that Gamelin sent his strategic reserve, including his best armoured units, off in the direction of Breda in the Netherlands, and that Allied air units were not used to bomb the longest traffic jam in history, which stretched from the Rhine through the Ardennes to the Meuse.[16] Some authorities link Gamelin's mistakes to French operational doctrine and the defensive strategic stance of the Phoney War.[17] But writers on military affairs tend to underline the importance of human agency at command level, and it might seem somewhat arbitrary to assign responsibility for Gamelin's false moves to an impersonal doctrine or mindset. If we remain at the same analytical level and remove doctrine as the decisive variable, German victory could be attributed to the energetic exploitation of the opportunities that arose as the campaign unfolded, whereas French defeat could be ascribed to the egregious mistakes of the Commander-in-Chief.

The isolation of the operational level

But can analysis be restricted to the operational level? Comparing operational doctrines begs the question whether one can justify isolating such a comparison when two such different societies were fighting each other as the totalitarian, Nazi *Führerstaat* and an alliance of liberal, parliamentary democracies.[18] Perhaps victory in 1940 is not the right criterion by which to judge the efficacy of Blitzkrieg as doctrine. After all, for all its putative operational brilliance Germany lost not one, but two world wars – a salient fact which is strangely absent from RMA literature. Perhaps German doctrine even contributed towards those defeats. The German High Command's concentration on operational issues to the exclusion of economic or other aspects of grand strategy certainly did. In wars between industrialized societies, the operational level has to be interpreted within the broader context of the resources available for total war.

Perhaps the French doctrine of *bataille conduite*, or methodical battle, contributed towards the final Allied victory. It has recently been described as representing a necessary adaptation of mass conscript armies to the conditions of industrialized warfare.[19] It was, after all, designed to mobilize superior material resources in a long war and thus suited the wait-and-see strategic stance of the Phoney War that was to allow the British and French to marshal their enormous resources and win the arms race. The danger of an enemy breakthrough was to be countered by a strategic reserve comprising the best armoured units and the rapid build-up of a modern air force.[20] The German way of war could better be described as a race against time to break out of the constraints of its restricted resource base and overheating economy.[21] It was less the expression of a superior operational doctrine than of a tactical training which encouraged aggression and initiative, in combination with an expansionist ideology and resource constraints, both of which pushed in the direction of unlimited conquest.

The isolation of the campaign against France

Certainly, if the object of the exercise is to study the efficacy of Blitzkrieg as an operational doctrine when put into practice, then the French campaign of 1940 is not the best historical example. As Karl-Heinz Frieser has put it, emphatically: '*The campaign in the West in 1940 was an unplanned but successful "Blitzkrieg", whereas the campaign in the East in 1941 was a planned but unsuccessful "Blitzkrieg"*.'[22] Or, in Wilhelm Deist's words: 'The attack on the Soviet Union, Operation BARBAROSSA, was actually the only German military effort in the Second World War planned as a Blitzkrieg campaign to be completed within limited time and with limited forces. In accordance with Hitler's aims, it was also planned as a racist war of annihilation.'[23]

In other words, if you want to study Blitzkrieg, you have to study 'Barbarossa'. But the Eastern Front gets scant attention in RMA literature. There would seem to be two reasons for this. First, despite Stalin's obliging mistakes and the lack of preparation on the part of the Red Army, the staggering initial success of the Wehrmacht did not bring victory by the end of the year. Blitzkrieg failed at the operational level because the infantry and logistics could not keep up with the advance of the Panzer divisions. At the level of grand strategy (which can hardly be said to have existed in Hitler's Germany) it failed because the arms output of the economy was not coordinated with the needs of the Wehrmacht. The OKW concentrated on its operational tasks and did not seek to secure the resources it needed even to fight a limited campaign.[24] The High Command thus bore a major share of the responsibility for the failure of its own operational concept.

It also bore a major share of responsibility for the crimes committed on the Eastern Front. This is a probable second reason why Blitzkrieg fans concentrate on the Western campaign. The very same officer corps, whose professional ethos is so much admired, built the open-air camps in which three million Soviet prisoners of war died for lack of food, shelter and medical attention.[25] They faithfully carried out the order to shoot several thousand political commissars.[26] They committed many atrocities during the course of anti-partisan warfare, which was also inextricably caught up in the destruction of Jews and Gypsies on Soviet territory.[27] Had the Blitzkrieg resulted in victory in 1941 it would not only have caused the downfall of communism, it would have been followed immediately by the worst genocide ever planned in history: the starving to death of some 30 million people in the north-west of the Soviet Union, who were to be cut off from the agricultural surplus in the south.[28] The decision to implement the Holocaust that *was* carried out, has itself been linked to the failure of the Blitzkrieg in the East.[29]

Nazi warfare

Several of these objectives were of course ideologically determined by the leadership of the Nazi state; but there is no doubt that the world view of the upper echelons of the German officer corps overlapped with Nazism's hatred of communism, democracy, world Jewry and the Slavic peoples.[30] It would require very tricky brain surgery to excise the professional ethos necessary to wage Blitzkrieg from the militaristic world view which waged genocidal warfare.[31] Indeed, the dovetailing objectives cannot be interpreted as the ideological pollution of professional military thought. Rather, it was the fundamental tenets of German military thought which produced both what is described as Blitzkrieg and what is understood by Nazi warfare: the unquestioned acceptance of war as the necessary arbiter in international affairs, the advocacy of total mobilization for total war, the abdication of responsibility for economic coordination and the concomitant concentration on the operational level of warfare, the integration of operational strategy in campaigns of conquest and plunder. Blitzkrieg was not an alternative to total war, but an expression – or rather the spearhead – of total war.[32]

If there was a degree of coherence in Nazi grand strategy, it was perhaps dictated by economic constraints: because Germany lacked the resource base to prevail in a long drawn-out struggle with the economic might of the British and French empires, the United States and the Soviet Union, it had to conquer new territories quickly and plunder their resources to maintain its momentum in further campaigns of conquest. To prevent the collapse of civilian morale, as had happened in 1918, slave labour, food and other raw materials were funnelled back to the Reich to ameliorate the rigours of war.[33] As Götz Aly has recently demonstrated, each individual soldier was given an artificial purchasing power that turned the Wehrmacht into a swarm of locusts all over occupied Europe.[34] Military necessity dovetailed neatly with Nazi expansionism and it is pointless – or missing the point – to isolate a sphere of operations full of useful lessons for today from the plunder, the massacres, the oppression and the genocide of the occupation regimes. Nazi Germany was not defeated in spite of its operational successes; it was the military thought that produced Blitzkrieg, which also led Germany to defeat.

The presentation of Blitzkrieg given above, pieced together from the work of scholarly authorities in the field, is merely intended to represent an alternative permutation of elements which are not taken into consideration in the interpretation of Blitzkrieg as an RMA. Squaring them off against each other would entail a discussion of the relative merits of traditional military history as opposed to the 'war and society' approach, along with other methodological and empirical issues familiar to historians from decades of debate. That has not been the object of the exercise in constructing this alternative. It is rather meant to be used as a foil against which the unspoken assumptions of defense intellectuals become visible.

The Blitzkrieg of the RMA literature, and of manoeuvre warfare before that,[35] is really an American ideal projected backwards into history, and then read out of history as the precursor and precedent of the way in which the United States should preferably wage war. As such, it is an example of the methodological trap into which defense intellectuals occasionally fall. The unspoken assumptions that lead them towards the trap are, first, that war is decided at the operational level by daring generals leading military machines that weld together soldiers and technologically advanced weaponry by means of cohesive doctrine. The moral seems to be: give the generals their head and there will be little collateral damage, no long drawn-out attrition, no need to mobilize the economy or respond to fickle public opinion.

Furthermore, it can be assumed that regime change is easy: it only takes six weeks for an army prepared for Blitzkrieg. Perhaps the conclusion is also drawn, on historical analogy, that the subsequent occupation will be self-financing and entail a modicum of garrison duty, but little of the partisan warfare and reprisals against civilians that can make post-invasion stabilization so messy. A compliant general can perhaps be wheeled out of retirement to take charge of a quiescent population. It is difficult to know how far the analogy with France in 1940 is assumed to go. There is no speculation on the follow-up to victory in the RMA literature, but perhaps the concentration on the operational means to winning a war quickly, coupled with an aversion to the realities of the eastern Blitzkrieg after 1941, serve the purpose of making war seem easier, more controllable, more affordable, more applicable.

Other assumptions about the symmetrical nature of conflict may derive from the lack of any credible threat to the United States during the 1990s. Some military professionals have criticized them as a tendency to let capabilities determine strategy in an environment devoid of any realistic scenario.[36] Yet they also testify to the hold that the 1940 French campaign has on the imagination. This is how war preferably should be fought, both in the past and in the future, the default option of military bureaucracies.

These unspoken assumptions probably reflect the post-Vietnam reorientation of U.S. military thought towards the operational level of war that Hew Strachan has identified.[37] They could also be a specific product of the rarefied atmosphere of the think-tanks, but due to the dearth of studies this can only be the subject of speculation.[38] It is hard to explain why the defense intellectual *milieu* has received so little academic attention when it plays so obvious a role in forming American strategic culture and exerts an undeniable influence over both foreign and domestic policy.[39] It must represent one of the most powerful, unelected groups within the American polity, and the purposes served by its research should presumably be the subject of public interest and scrutiny. In any other field, it is assumed – or suspected – that the products of research institutions are influenced by funding, political affiliations, institutional rivalry and individual career structures. In this one, however, their impact on theory can only be guessed at. One insider has compared the quick succession of High Concepts in American strategic debate to the workings of the fashion industry.[40] If that is a valid comparison, it is also justifiable to ask what mechanisms exist within the industry to weed out the bad theory that fads and market mechanisms will inevitably produce.

There is also a third avenue that could be explored in future studies of the American image of German warfare. There does seem to exist a longstanding and uncritical admiration for the Wehrmacht in defense intellectual circles. Samuel Huntington's *The Soldier and the State* is regarded as a classic text; yet it contains some very strange claims about the Reichswehr and the Wehrmacht, intended to demonstrate that the impeccable professionalism of the Prussian–German officer corps had been corrupted by extraneous political influences.[41]

In his admiration for the German officer corps Huntington would not recognize that its military culture had fostered an understanding of operational 'necessity' with disastrous political consequences. This led to calls for preventive war before 1914 and to the militarization of state, society and economy during World War One. The lessons drawn by professional military thought from that conflict overlapped with the world view of Nazism; it was not polluted by outside influences, as Huntington's ideal type asserted it must have been. The Reichswehr worked to replace the democratic Weimar constitution by an authoritarian government that could better prepare the total mobilization needed to fight total war, and the Wehrmacht worked hand in glove with Hitler's regime to prepare for it.[42]

Huntingon, however, denied that Nazi and military values overlapped in any way. This was an untenable position, even in 1957.[43] He ignored the Nuremberg trial of the OKW and in effect parroted the self-exculpatory, post-war writings of the Wehrmacht leadership. In so doing he was following in the footsteps of the U.S. Army's Historical Division, which had provided 300 German officers with the opportunity to write their own operational history.[44]

That one of the worst examples of destructive militarism can be portrayed as a paragon of military professionalism, suggests that there is something wrong with Huntington's theory of civil–military relations. That he is still regarded as a classical thinker within some disciplines, suggests that his admiration for the Wehrmacht is common currency, or that they lack a culture of scholarly criticism, or both.[45] To the extent that reverence for German military achievements continues in defense intellectual circles, it serves only to perpetuate the post-war apologia of the Wehrmacht generals at the very moment the tissue of fabrications upon which they rest is being unravelled by German historians.[46] If this is symptomatic of the lack of a critical academic culture, admiration for the Wehrmacht and Blitzkrieg may not be the only assumptions in need of revision.

Notes

1 The 's' in 'defense' acknowledges American dominance in the field. I define defense intellectuals as academically trained theorists who circulate among think-tanks, branching off in the course of their careers into the military, the arms and security industries, universities or administrations. Their work draws heavily on the classics of military theory, uses historical illustrations and applies political science models. Whether or not this specific mix of disciplines is itself *disciplined*, in the sense of being subjected to a rigorous methodology designed to weed out bad theory, is a question this chapter seeks to raise. Given the astounding lack of studies of the think-tank universe, it can probably not be answered at present.

2 (For reasons of space, references throughout the chapter are limited to significant studies or syntheses.) Andrew F. Krepinevich, *The Military-Technical Revolution: A Preliminary Assessment*, Washington, D.C.: Center for Strategic and Budgetary Assessments, 2002 (written 1992), pp. 5, 30; Andrew F. Krepinevich, 'Cavalry to Computer: The Pattern of Military Revolutions', *The National Interest*, 1994, pp. 30–42: 37; James R. Fitzsimmonds and Jan M. Van Tol, 'Revolutions in Military Affairs' in *Joint Force Quarterly*, Spring 1994, pp. 24–31: 24f.; Eliot A. Cohen, 'A Revolution in Warfare' in *Foreign Affairs* 75: 2, 1996, pp. 37–54: p. 46; Williamson Murray, 'Armoured Warfare: The British, French and

German Experiences' in Williamson Murray and Allan R. Millett (eds) *Military Innovation in the Interwar Period*, Cambridge: Cambridge University Press, 1996, pp. 6–49: p. 40; Williamson Murray, 'Thinking About Revolutions in Military Affairs' in *Joint Force Quarterly*, Summer 1997, pp. 69–76, 70; Richard O. Hundley, *Past Revolutions, Future Transformations: What Can the History of Revolutions in Military Affairs Tell Us About Transforming the U.S. Military?*, Santa Monica, CA: Rand, 1999, p. 11; Clifford J. Rogers, '"Military Revolutions" and "Revolutions in Military Affairs"' in Thierry Gongora and Harald Von Riekhoff (eds), *Toward a Revolution in Military Affairs?: Defense and Security at the Dawn of the Twenty-First Century*, Westport, Conn.: Greenwood Press, 2000, pp. 21–35, p. 23; Williamson Murray, 'May 1940: Contingency and Fragility of the German RMA', in MacGregor Knox and Williamson Murray (eds), *The Dynamics of Military Revolution, 1300–2050*, Cambridge: Cambridge University Press, 2001, pp. 154–74, 155; Max Boot, *War Made New: Technology, Warfare, and the Course of History, 1500 to Today*, New York: Gotham Books, 2006, pp. 13, 234.

3 Larry H. Addington, *The Blitzkrieg Era and the German General Staff, 1865*–1941, New Brunswick, N.J.: Rutgers University Press, 1971, p. 122; Trevor N. Dupuy, *A Genius for War: The German Army and General Staff, 1807–1945*, Englewood Cliffs, N.J.: Prentice-Hall, 1977, pp. 212–15, 237–40, 255ff.; Edward N. Luttwak, 'The Operational Level of War' in *International Security* 5:3, 1981, pp. 61–79; Barry Posen, *The Sources of Military Doctrine: France, Britain, and Germany between the World Wars*, Cornell Studies in Security Affairs, Ithaca: Cornell University Press, 1984, pp. 205–19; Williamson Murray, 'German Army Doctrine, 1918–1939, and the Post-1945 Theory Of "Blitzkrieg Strategy"', in Carole Fink *et al.* (eds), *German Nationalism and the European Response, 1890–1945*, Norman: University of Oklahoma Press, 1985, pp. 71–94; James S. Corum, *The Roots of Blitzkrieg: Hans Von Seeckt and German Military Reform*, Lawrence, Kan.: University Press of Kansas, 1992; Robert Michael Citino, *The Path to Blitzkrieg: Doctrine and Training in the German Army, 1920–1939*, Boulder, Colo.: Lynne Rienner Publishers, 1999; Mary R. Habeck, *Storm of Steel: The Development of Armor Doctrine in Germany and the Soviet Union, 1919–1939*, Ithaca, N.Y.: Cornell University Press, 2003; Robert Michael Citino, *Blitzkrieg to Desert Storm: The Evolution of Operational Warfare*, Lawrence, Kan.: University Press of Kansas, 2004.

4 Matthew Cooper, *The German Army, 1933–1945: Its Political and Military Failure*, London: Macdonald and Jane's, 1978, pp. 113–66. Cooper describes in detail the growth of the 'armoured idea' before the war but emphatically rejects the notion that any revolutionary theory of Blitzkrieg supplanted the traditional strategy of envelopment and annihilation. Michael Geyer, 'German Strategy in the Age of Machine Warfare, 1914–1945', in Peter Paret (ed.), *Makers of Modern Strategy: from Machiavelli to the Nuclear Age*, Princeton, N.J.: Princeton University Press, 1986, pp. 527–97: 585f.; J.P. Harris, 'The Myth of Blitzkrieg' in *War in History* 2: 3, 1995, pp. 335–52; Karl-Heinz Frieser, *Blitzkrieg-Legende: Der Westfeldzug 1940*, Munich: R. Oldenbourg, 1996. Frieser's study is fundamental and was published in English translation in 2005. Azar Gat, *British Armour Theory and the Rise of the Panzer Arm: Revising the Revisionists*, London: Macmillan Press, 2000, p. 83f.; Azar Gat, 'Ideology, National Policy, Technology and Strategic Doctrine between the World Wars' in *Journal of Strategic Studies* 24:3, 2001, pp. 1–18: p. 10; Denis E. Showalter, 'More Than Nuts and Bolts: Technology and the German Army' in *The Historian* 65, Fall 2002, pp. 123–43; Gil-li Vardi, 'The Enigma of German Operational Theory: The Evolution of Military Thought in Germany, 1919–1938', Ph.D. dissertation, The London School of Economics and Political Science, 2008.

5 Raymond Aron, *Penser la Guerre, Clausewitz*, Paris: Gallimard, 1976, Vol. I, pp. 122–42, 412–24; Arden Bucholz, *Hans Delbrück & the German Military Establishment: War Images in Conflict*, Iowa City: University of Iowa Press, 1985; Azar Gat,

A History of Military Thought: from the Enlightenment to the Cold War, Oxford: Oxford University Press, 2001, pp. 371–7.

6 This is the approach adopted by Stephen Biddle in his critique of Blitzkrieg as a putative RMA. Stephen Biddle, 'The Past as Prologue: Assessing Theories of Future Warfare' in *Security Studies* 8:1, 1998, pp. 1–74, 51.

7 Frieser, *Blitzkrieg-Legende*, p. 117.

8 Karl-Volker Neugebauer, 'Operatives Denken zwischen dem Ersten und Zweiten Weltkrieg' in *Operatives Denken und Handeln in Deutschen Streitkräften im 19. und 20. Jahrhundert*, Herford, Bonn: E.S. Mittler & Sohn, 1988, pp. 97–122.

9 Gerhard P. Gross, 'Das Dogma der Beweglichkeit. Überlegungen zur Genese der deutschen Heerestaktik im Zeitalter der Weltkriege' in Bruno Thoss and Hans Erich Volkmann (eds.), *Erster Weltkrieg, Zweiter Weltkrieg: Ein Vergleich: Krieg, Kriegserlebnis, Kriegserfahrung in Deutschland*, Paderborn: F. Schöningh, 2002, pp. 143–66: 153f.

10 Cf. Gat, *British Armour Theory* and R.L. DiNardo, 'German Armour Doctrine: Correcting the Myths' in *War in History* 3:4, 1996, pp. 384–97.

11 Vardi, *German Operational Theory*, p. 183.

12 F.O. Miksche, *Blitzkrieg*, London: Faber and Faber, 1941. For the permutations of the term, see George Raudzens, 'Blitzkrieg Ambiguities: Doubtful Usage of a Famous Word' in *War and Society* 7:2, 1989, pp. 77–94.

13 Quoted in Karl-Heinz Frieser, 'Die Deutschen Blitzkriege: Operativer Triumph – Strategische Tragödie' in Rolf-Dieter Müller and Hans-Erich Volkmann (eds.), *Die Wehrmacht: Mythos und Realität*, München: Oldenbourg, 1999, pp. 182–96, p. 192.

14 For a survey of recent research, see Joel Blatt, *The French Defeat of 1940: Reassessments*, Providence, R.I.: Berghahn Books, 1998. Cf. Frieser, *Blitzkrieg-Legende* and Ernest R. May, *Strange Victory: Hitler's Conquest of France*, New York: Hill and Wang, 2000.

15 Denis Showalter, 'Ce que l'armée française avait compris de la guerre moderne', in Maurice Vaïsse (ed.), *Mai-Juin 1940. Défaite française, victoire allemande, sous l'œuil des historiens étrangers*, Paris: Autrement, 2000, pp. 29–58, 38.

16 Don Alexander, 'Repercussions of the Breda Variant' in *French Historical Studies* 8, 1974, pp. 459–88; Jeffrey A. Gunsburg, 'Coupable ou non? Le rôle du Général Gamelin dans la défaite de 1940' in *Revue historique des armées* 4, 1979, pp. 145–63; Frieser, *Blitzkrieg-Legende*, pp. 106–10, p. 135; Showalter, 'Armée française', p. 56.

17 E.g. Philippe Garraud, 'L'idéologie de la "défensive" et ses effets stratégiques: le rôle de la dimension cognitive dans la défaite de 1940' in *Revue francaise de science politique* 54, 2004, pp. 781–810: pp. 796, 804f.

18 A similar point with regard to the comparison between German and Soviet doctrine is made in Gat, 'Ideology', p. 13f.

19 Showalter, 'Armée française', 54f.; Douglas Porch, 'Military "Culture" and the Fall of France in 1940' in *International Security* 24, 2000, pp. 157–80.

20 Gat, *British Armour Theory*, pp. 27–30.

21 Discussion of the economic pressures and constraints on Blitzkrieg has continued for more than 40 years. A few stepping stones are Alan S. Milward, *The German Economy at War*, London: Athlone Press, 1965, pp. 1–27; Timothy W. Mason, 'Internal Crisis and War of Aggression, 1938–1939' in Jane Caplan (ed.), *Nazism, Fascism and the Working Class*, Cambridge: Cambridge University Press, 1995, pp. 104–30; R. J. Overy, *War and Economy in the Third Reich*, Oxford: Clarendon Press, 1994, pp. 177–256; Adam Tooze, *The Wages of Destruction: The Making and Breaking of the Nazi Economy*, London: Allen Lane, 2006.

22 Frieser, *Blitzkrieg-Legende*, p. 439. (Frieser's emphasis).

23 Wilhelm Deist, 'The Road to Ideological War: Germany, 1918–1945' in MacGregor

Knox *et al.*, *The Making of Strategy: Rulers, States, and War*, Cambridge: Cambridge University Press, 1994, pp. 352–92, p. 389. Also Bernhard R. Kroener, 'Der "erfrorene" Blitzkrieg. Strategische Planungen der deutschen Führung gegen die Sowjetunion und die Ursachen ihres Scheiterns' in Bernd Wegner (ed.), *Zwei Wege nach Moskau: Vom Hitler-Stalin-Pakt bis zum 'Unternehmen Barbarossa'*, München: Piper, 1991, pp. 133–48, 145.

24 Wilhelm Deist, '"*Blitzkrieg*" Or Total War? War Preparations in Nazi Germany', in Roger Chickering and Stig Förster (eds), *The Shadows of Total War: Europe, East Asia, and the United States, 1919–1939*, Cambridge: Cambridge University Press, 2003, pp. 271–84.

25 Christian Streit, *Keine Kameraden: Die Wehrmacht und die sowjetischen Kriegsgefangenen 1941–1945*, Stuttgart: Deutsche Verlags-Anstalt, 1978.

26 Felix Römer, *Der Kommissarbefehl: Wehrmacht und NS-Verbrechen an der Ostfront 1941/42*, Paderborn: Schöningh, 2008. The war in the east has been the subject of a major research programme recently concluded by the Institut für Zeitgeschichte, see Johannes Hürter, *Hitlers Heerführer: Die deutschen Oberbefehlshaber im Krieg gegen die Sowjetunion 1941/42*, Munich: Oldenbourg, 2006; Dieter Pohl, *Die Herrschaft der Wehrmacht: Deutsche Militärbesatzung und einheimische Bevölkerung in der Sowjetunion 1941–1944*, Munich: Oldenbourg, 2008; Christian Hartmann, *Wehrmacht im Ostkrieg: Front und militärisches Hinterland 1941/42*, Munich: Oldenbourg, 2009.

27 Hannes Heer and Klaus Naumann (eds), *Vernichtungskrieg: Verbrechen der Wehrmacht 1941–1944*, Hamburg: Hamburger Edition, 1995; Jürgen Förster, 'Wehrmacht, Krieg und Holocaust' in Müller and Volkmann, *Die Wehrmacht*, pp. 948–63.

28 Rolf Hobson, *Krig og strategisk tenkning i Europa 1500–1945: samfunnsendring – statssystem – militær teori*, Oslo: Cappelen, 2005, p. 296.

29 Omer Bartov, 'From *Blitzkrieg* to Total War: Controversial Links between Image and Reality' in Ian Kershaw and Moshe Lewin, *Stalinism and Nazism: Dictatorships in Comparison*, Cambridge: Cambridge University Press, 1997, pp. 158–184: 174ff.; cf. Tobias Jersak, 'Blitzkrieg Revisited: a New Look at Nazi War and Extermination Planning' in *The Historical Journal* 43, 2000, pp. 565–82

30 Wolfram Wette, *Die Wehrmacht: Feindbilder, Vernichtungskrieg, Legenden*, Frankfurt am Main: S. Fischer, 2002. For a more nuanced exposition of German views of Russia, see Gerd Koenen, *Der Russland-Komplex: Die Deutschen und der Osten 1900–1945*, München: Beck, 2005.

31 Aspiring brain surgeons now have the opportunity to practise on one of the heroes of RMA literature, Erich von Manstein, thanks to Oliver von Wrochem, *Erich Von Manstein: Vernichtungskrieg und Geschichtspolitik*, Paderborn: F. Schöningh, 2006. Unfortunately, the greatest hero, Guderian, has not yet received the same treatment.

32 Hew Strachan, *European Armies and the Conduct of War*, London; Boston: Allen & Unwin, 1983, p. 163f.

33 Ulrich Herbert, *Hitler's Foreign Workers: Enforced Foreign Labor in Germany under the Third Reich*, Cambridge: Cambridge University Press, 1997. The Nazi leadership actually believed in the 'stab in the back' legend which did so much to bring them to power; the loyalty of the German population had therefore to be preserved by plundering the resources of others. For the reality of societal dissolution *and* military collapse in 1918, and the widespread belief that only the former had brought about surrender, see Boris Barth, *Dolchstosslegenden und politische Desintegration: Das Trauma der deutschen Niederlage im Ersten Weltkrieg 1914–1933*, Düsseldorf: Droste, 2003.

34 Götz Aly, *Hitlers Volksstaat: Raub, Rassenkrieg und Nationaler Sozialismus*, Frankfurt am Main: S. Fischer, 2005.

35 Asa A. Clark, *The Defense Reform Debate: Issues and Analysis*, Baltimore: Johns Hopkins University Press, 1984; William S. Lind, *Maneuver Warfare Handbook*, Boulder, Colo.: Westview Press, 1985; Robert R. Leonhard, *The Art of Maneuver: Maneuver-Warfare Theory and Airland Battle*, Novato, CA: Presidio Press, 1991; Richard D. Hooker, *Maneuver Warfare: An Anthology*, Novato, CA: Presidio Press, 1993.

36 Vincent Desportes, *La guerre probable: penser autrement*, Paris: Economica, 2007, pp. 100–14; H.R. McMaster, 'On War: Lessons to Be Learned' in *Survival* 50, 2008, pp. 19–30.

37 Hew Strachan, 'The Lost Meaning of Strategy' in *Survival* 47, 2005, pp. 33–54; Hew Strachan, 'Civil–Military Relations after Iraq' in *Survival* 48, 2006, pp. 59–82, 60f.

38 Twenty-five years separate two important studies of RAND, Fred M. Kaplan, *The Wizards of Armageddon*, New York: Simon and Schuster, 1983; Alex Abella, *Soldiers of Reason: The Rand Corporation and the Rise of the American Empire*, Orlando: Harcourt, Inc., 2008.

39 An exception is Dima Adamsky, *The Culture of Military Innovation: the Impact of Cultural Factors on the Revolution in Military Affairs in Russia, the U.S., and Israel*, Stanford, Calif.: Stanford University Press, 2010.

40 Colin S. Gray, *Strategy for Chaos: Revolutions in Military Affairs and the Evidence of History*, London: Frank Cass, 2002, p. 17f.

41 Samuel P. Huntington, *The Soldier and the State; the Theory and Politics of Civil–Military Relations*, Cambridge: Belknap Press of Harvard University Press, 1957, pp. 99–124.

42 In lieu of a vast literature, a succinct synthesis can be found in Klaus-Jürgen Müller, 'Die Reichswehr und die "Machtergreifung"', in Wolfgang Michalka (ed.), *Die nationalsozialistische Machtergreifung*, Paderborn, 1984, pp. 137–51.

43 Three fundamental studies published in the 1950s were John Wheeler-Bennett, *The Nemesis of Power; the German Army in Politics, 1918–1945*, London, New York: Macmillan: St. Martin's Press, 1953; Gerhard Ritter, *Staatskunst und Kriegshandwerk; das Problem des "Militarismus" in Deutschland*, Munich: R. Oldenbourg, 1954, Vol. I; and Gordon Alexander Craig, *The Politics of the Prussian Army 1640–1945*, Oxford: Clarendon Press, 1955. Although Huntington quoted them, he left out the cautionary tale about Prussia being 'an army with a state', as well as the deleterious political consequences of the General Staff's war planning and the militarization of state and society, and concentrated on the politicization of the military. On the Reichswehr's contribution to the destruction of democracy, he could also have quoted Karl Dietrich Bracher, *Die Auflösung Der Weimarer Republik: Eine Studie zum Problem des Machtverfalls in der Demokratie*, Stuttgart: Ring-Verlag, 1955.

44 Bernd Wegner, 'Erschriebene Siege. Franz Halder, die "Historical Division" und die Rekonstruktion des Zweiten Weltkrieges im Geiste des deutschen Generalstabes', in Ernst Willi Hansen *et al.* (eds), *Politischer Wandel, organisierte Gewalt und nationale Sicherheit: Beiträge zur neueren Geschichte Deutschlands und Frankreichs: Festschrift für Klaus-Jürgen Müller*, Munich: Oldenbourg, 1995, pp. 287–304.

45 There is a nascent awareness of the problematic aspect of Huntington's treatment of Germany in Williamson Murray, 'Professionalism and Professional Military Education in the Twenty-first Century' in Suzanne C. Nielsen and Don M. Snider (eds), *American Civil–Military Relations: The Soldier and the State in a New Era*, Baltimore: Johns Hopkins University Press, 2009, pp. 133–48, p. 134. But it should be noted that Huntington chose to ignore not only the Wehrmacht's crimes against humanity, but also its crimes against democracy.

46 Kurt Pätzold, *Ihr waret die besten Soldaten: Ursprung und Geschichte einer Legende*, Leipzig: Militzke, 2000.

12 Conclusion

Military innovation between anticipation and adaptation

Dmitry (Dima) Adamsky and Kjell Inge Bjerga

Innovations in military regimes are inevitable and, as pointed out by Azar Gat, some of them become the defining moments of modernity. This book concentrated on contemporary military thought and innovation and explored several aspects of the Information Technology Revolution in Military Affairs (IT-RMA). The book described how this innovation was incepted, generated and emulated, what counter-measures it faced, and how the IT-RMA-type militaries adapted their thinking and fighting to their respective operational challenges.

The authors demonstrated that modern militaries innovate, and contemporary military theory develops, through two complementary processes. One process is when military theoreticians imagine the future of warfare and outline new theories of victory. Such *anticipation* emerges in a deductive manner and innovation evolves in the top-down mode. Anticipation is primary a peacetime effort that involves a lot of imagination, systematic exploration and holistic grasp of the current character of war. The second process is a bottom-up innovation, which is more inductive and reactive in its nature and is thus dubbed *adaptation*. Adaptation is primary a wartime effort that builds on the insights produced by battleground friction and on the lessons learned from the best practices. At the outset the IT-RMA has been indeed a story of the top-down innovation that was incepted as a peacetime endeavor. It evolved, however, during the wartime, as a bottom-up adaptation.

The book showed that the development of military thought is a dialectical process. Deductive military thinkers outline a new theory of victory. When this thesis meets the antithesis of the real war, adaptation comes into play. Eventually, the best practices of the successful adapters find their way to the higher echelons of doctrine writers, and become codified as the postulates of the military theory, until this theory, in its turn, meets the new round of operational reality check. The chapters illuminated how this dialectical process took place in the context of the IT-RMA. In keeping with the theoretical overviews, outlined by Azar Gat and Harald Høiback, empirical chapters by Stephen P. Rosen and Antulio J. Echevarria demonstrated that the notion of the IT-RMA was born and conceptualized as a deductive, peacetime, top-down innovation. While Rosen concentrated on the inception of the IT-RMA vision, Echevarria showed how this generic theoretical notion was transformed into concrete doctrinal outline for the U.S. armed forces. Kjell Inge Bjerga and Torunn Laugen Haaland demonstrated a similar tendency for top-down, peacetime, innovation in the armed forces of the small state, and Jacqueline Newmyer Deal traced the same occurrence in the military of the great power exploring the Chinese approach to the IT-RMA.

However, as Itai Brun and Carmit Valensi showed, while the Western armed forces were busy outlining and diffusing their thesis of future war, a parallel school of military thought originated within several state and non-state strategic communities in the Middle East. Their thinking about the future of warfare progressed in a different direction. When an encounter between these two streams of intellectual energy took place on the actual battleground, an acute need emerged among the IT-RMA militaries to adjust the models of operational behavior and doctrinal thinking. Theo Farrell and James Russell showed in their chapters how operational friction produces an antithesis for the idealized IT-RMA. They demonstrated how tactical *adaptation* of the U.S. and British militaries bridged the gap between several theoretical postulates and operational reality on the ground. Innovative best practices eventually produced through adaptation evolved into a mainstream doctrinal thinking.

Each of these two approaches to challenges of warfare and to military theory development – *anticipation* and *adaptation* – has its pros and cons. *Anticipation* implies institutional and intellectual propensity to imagine the most probable future. This aptitude enables the preparation of armed forces in advance for the prospective military regime. *Anticipation,* however, implies a great number of mistakes. Since uncertainty cannot be eliminated, this vision is almost certain to be vague and incomplete, rather than detailed, precise, or predictive.[1] In contrast, *adaptation* implies that one tolerates the possibility of surprise and accepts in advance a potential mismatch between one's military architecture and the demands of the future battlefield. Ability to innovate through *adaptation* enables a bridging of the gap between the imagined war and the real one.

Strategic interaction is the constant competition of learning among the adversaries and a competition of conceptual imagination and operational ingenuity. Success in the future battle depends equally on the ability to grasp the current character of war, to develop a 'new theory of victory' corresponding to it, as it does on the ability to adjust oneself to unforeseen developments once in a war situation. Thus, *anticipation* and *adaptation* are mutually reinforcing virtues. Since *adaptation* may be a more costly endeavor than *anticipation,* the one who prevails in the peacetime competition for anticipation will have to invest less in the wartime adjustments.

Harmonic orchestration of these efforts is likely to result in an increased military effectiveness. Consequently, preserving a healthy balance between these two types of innovation is essential. As this book showed, going to either extreme may lead to devastating consequences. Overemphasis on military futurology is counterproductive. Campaigns of the last decade forced those militaries that transformed themselves along the lines of the imagined IT-RMA models, to modify their visions of war and adjust them to the current security environment. Fascination with military futurology may result in wishful thinking. Indeed, several proponents of the IT-RMA outlined how they 'would like to fight and then assumed that the preference was relevant'. This self-delusion, of course, corrupted the ability to think critically about warfare.[2]

At the same time, insufficient investment in anticipating the future could be equally damaging. Military institutions constantly need to invest significant intellectual energy in developing a vision of a future war.[3] The fact that oracle forecasts in military affairs are problematic should not dissuade analysts from intellectual endeavors. Although fascination with imagining a 'next war' might turn into a dangerous professional pathology, preparations for 'this war' and thinking about it have to start before 'this war' actually needs to be fought. Otherwise a military organization might become irrelevant for the current security environment. Even if one cannot predict the appearance of the black swan, imagining and anticipating its existence can be a contribution in itself. This book showed that systematic thinking about the changing character of war is possible and rewarding.

The dynamic relationship between *anticipation* and *adaptation* can be observed in the chapter by Antulio J. Echevarria, who discusses the complex interaction between operational experiences and the evolution of the military theory and doctrine in the U.S. military services. Theo Farrell and James Russell also looked into interplay of RMA theory with actual operational practice and examined how the bottom-up wartime learning process (on the battlefield) in Iraq and Afghanistan adjusted the preconceptions about modern warfare in the British and the U.S. militaries. One conclusion emanating from the book's findings is that the better an *anticipation* effort is, the smaller will be the volume of *adaptation* that is needed to bridge the gap between the demands of the battleground and the characteristics of the fighting force.

During the 1990s IT-RMA militaries innovated on the most part through *anticipation*, which resulted in so-called defence transformation on both sides of the Atlantic. Since the mid-2000s, when most of the IT-RMA militaries found themselves on battlefields inconsistent with the idealized visions of future war, significant practical and intellectual energy was invested into innovation by *adaptation*. It might be that these days the time is ripe again for the next round of the *anticipation*. The focus on current wars and innovation through *adaptation* should not come at the expense of the systematic thinking about the future challenges and the emerging military regimes. Lack of vision about changes in the character of war is as dangerous as wishful thinking about it. Without the intellectual–institutional commitment to craft a vision of future war and the simultaneous commitment to constantly and critically engage the new paradigm, the military organization will almost certainly lose its relevance for any given security regime.

How can conduct and theories of warfare change in the foreseeable future? Although Rolf Hobson's chapter makes a reasonable methodological precaution for military theoreticians to carefully choose historical frames of reference when thinking about the future RMAs, questions pertaining to the future of warfare are beyond the scope and methodology of this book. Still, informed by the insights offered in this volume, we would like to outline some possible avenues for future research. One way to explore emerging discontinuities in military affairs would be to examine the prospective behavior of state and non-state actors in the context of precision, hybrid, nuclear and cyber warfare.

As Stephen P. Rosen argues in the conclusion of his chapter, at the time when the IT-RMA was still a peacetime innovation, initial *anticipation* efforts imagined 'large, technically capable adversaries, who attempted to use new capabilities to thwart older American power projection forces, and attacked the sources of information used by the United States to block American military objectives'. However, argues Rosen, the protracted and exigent campaigns of the last decade led to an urgent focus on the counterinsurgency related problems, and channeled the time and money of the military establishment away from the exploration of revolutionary change, as it was envisioned by the original RMA theoreticians. What if, in the near future, we are about to witness the revival of this trend? What if the precision strike becomes the dominant operational approach on both sides of the future battlefield?

This assumption is in keeping with an observation made by Gat that advanced hardware is not becoming prohibitively expensive but, on the contrary, turns very accessible. Indeed, accuracies and ranges improve incrementally around the world; relatively inexpensive guided weaponry proliferates to state and non-state actors, and advanced sensors have become more accessible worldwide. The 'O-RMA', outlined in the chapter by Itai Brun and Carmit Valensi, may also take a turn in this direction, as suggested by trends in warfare in the Middle East, the operational behavior of Hezbollah and Hamas during recent campaigns, and force build up of the Iranian military. A similar conclusion can be reached by reading Newmyer Deal's analysis of the 'anti-access, area denial' approach of the Chinese PLA. These symptoms, from around the world, suggest that we might be witnessing the early stages of the reconnaissance-strike complexes' proliferation, along the lines anticipated by the early IT-RMA theoreticians.

In what ways may the proliferations of long- and short-range precision-strike capabilities, advanced sensors and command-and-control assets transform the nature of the battlefield in the near future? In his recent work Barry Watts starts to explore this and other questions, examining the battlefield that features two adversaries armed with the reconnaissance-strike complexes. If assumptions about these emerging trends prove to be correct then the conduct of warfare between now and 2031 could change more significantly than between 1991 and now. If warfare indeed does shift in this direction, the military theoreticians may wish to explore new concepts of operations, organizational structures and force build-up. [4]

Another possible focus for the *anticipation* efforts is the nature of warfare in a 'proliferated world'. Today, security studies' scholars systematically explore a host of questions pertaining to potential expansion of the nuclear weapons club. Scrutiny focuses on sources and patterns of nuclear behavior among current and potential proliferators.[5] The exploration of the prospective nuclear battlefield may include the following questions: how would the newcomers to the nuclear club design their force and concept of operations? What involves operational planning in a multipolar nuclear environment? How would actors with different strategic cultures exploit these capabilities and how might their theories of victory differ? How will the emerging thinking about the nuclear weapons with tailored effects and tactical nuclear weapons influence global and regional powers' approach to warfare? These initial anticipation efforts should probably be further accelerated. In the 'proliferated world' the price of *adaptation* following the *anticipation* mistakes may be disproportionally high.

Another discussion about the changing character of war concentrates today on the questions pertaining to the hybrid and cyber battlefields. The hybrid, or compound warfare, refers to the battlefields where an adversary incorporates several strategic identities, as outlined in the chapter by Brun and Valensi, or when several forms of fighting activity, conventional, irregular and non-conventional, merge on the battleground, as observed by Frank Hoffman.[6] Anticipating trends of the hybrid warfare suggest several avenues for research. One example of such exploration is the question that Brun and Valensi raised with regard to the future of the O-RMA theory of victory resulting from the ramifications of the Arab Spring. Another example is the question raised by Newmyer Deal, who, elaborating on the Chinese understanding of the RMA, wonders to what extent Chinese RMA theorists integrate conventional and nuclear operations or distinguish sharply between them. This question may be extended beyond the regional context and applied to several other actors already operating on the existing and the emerging hybrid battlefields around the world.

Scholars of strategy and military innovations started to explore recently emerging cyber capabilities. Exploration so far focuses on examination of how this new tool of war influences such classical postulates of strategic theory as offense–defense and fire–maneuver balances. Anticipation efforts may focus on questions of whether this new tool of war increases or decreases strategic stability, what is the essence of deterrence in the cyber age, and whether it makes actors more risk-prone or risk-averse? How does it alter the character of warfare on the operational and tactical levels, and how does this shift, in its turn, impact on weapon systems, organizational structures and doctrine? What would be symmetrical and asymmetrical countermeasures to a cyber attack? What are the norms and taboos of warfare in the cyber age? The emergence of the 'cyber' has already affected the way we think about warfare, as it introduced the idea of 'war between the wars' (as opposed to the inter-warfare cyber attacks) transforming modern warfare from an episodic activity into an almost uninterrupted continuous campaign.

The history of the IT-RMA teaches us about the importance of balancing between military futurism on one hand and a necessary tool kit for ensuring organizational and conceptual flexibility and swift operational adaptation on the other. A military community would assure its innovative capacity if, while anticipating an imagined future war, it also cultivated an ability to be flexible enough to adapt if the initial anticipation was proved wrong.

Notes

1 For example see Barry Watts and Williamson Murray, 'Military Innovation in Peace Time', in Williamson Murray and Allan Millet, *Military Innovation in the Interwar Period*, Cambridge: Cambridge UP, 1996.
2 For example see: H. R. McMaster, 'On War: Lessons to be Learned' in *Survival*, 50:1, February 2008, p. 19, and 'Learning from Contemporary Conflicts to Prepare for Future War' in *Orbis*, 52:4, Fall 2008, pp. 564–84.
3 Watts and Murray, 'Military Innovation in Peace Time', pp. 404–6.
4 Barry D. Watts, *The Maturing Revolution in Military Affairs*, Washington D.C.: CSBA, 2011.
5 For example see Michael Brown *et al.*, *Going Nuclear: Nuclear Proliferation in the 21st Century*, Cambridge, Mass.: The MIT Press, 2010; William Potter and Gaukhar Mukhtazhanova, *Forecasting Nuclear Proliferation in the 21st Century*, Stanford, Calif.: Stanford UP, 2010; Andrew W. Marshall, 'Strategy as a Profession' in Robert Zarate and Henry Sokoloski, *Nuclear Heuristics*, Carlisle, Pa.: Strategic Studies Institute, 2009.
6 Frank G. Hoffman, *Conflict in the 21st Century: the Rise of Hybrid Wars*, Arlington, Virginia: Potomac Institute for Policy Studies, 2007.

Bibliography

Abella, Alex, Soldiers of Reason: The Rand Corporation and the Rise of the American Empire, Orlando: Harcourt Inc., 2008.

Adamsky, Dima, 'American Strategic Culture and the U.S. Revolution in Military Affairs' in Defence and Security Studies, 1, 2008.

Adamsky, Dima, 'Through the Looking Glass: The Soviet Military–Technical Revolution and the American Revolution in Military Affairs' in Journal of Strategic Studies, 31:2, 2008, pp. 257–94.

Adamsky, Dima, 'Jihadi Operational Art: The Coming Wave of Jihadi Strategic Studies' in Studies in Conflict & Terrorism, 33:1, January 2010.

Adamsky, Dima, The Culture of Military Innovation: The Impact of Cultural Factors on the Revolution in Military Affairs in Russia, the U.S. and Israel, Stanford, CA: Stanford UP, 2010.

Addington, Larry H., The Blitzkrieg Era and the German General Staff, 1865–1941, New Brunswick, NJ: Rutgers University Press, 1971.

Al Suri, A., 'The Call to Global Islamic Resistance' in Jim Lacey, The Canons of Jihad: Terrorists' Strategy for Defeating America, Annapolis: Naval Institute Press, 2006.

Alberts, David S., John J. Gartska and Frederick P. Stein, *Network Centric Warfare: Developing and Leveraging Information Superiority*, Washington DC: C4ISR Cooperative Research Program, 1999.

Alderson, Alex, 'The Validity of British Army Counterinsurgency Doctrine after the War in Iraq, 2003–2009' in Defence Academy College of Management and Technology, Cranfield University, 2009.

Alexander, Don, 'Repercussions of the Breda Variant' in *French Historical Studies* 8, 1974, pp. 459–88.

Aly, Götz, *Hitlers Volksstaat: Raub, Rassenkrieg und Nationaler Sozialismus*, Frankfurt am Main: S. Fischer, 2005.

Aron, Raymond, *Penser la Guerre, Clausewitz*, 2 vols, Bibliothèque Des Sciences Humaines, Paris: Gallimard, 1976.

Avant, Deborah D., *Political Institutions and Military Change: Lessons from Peripheral Wars*, Cornell: University Press, 1994.

Baocun, Wang, 'Military Reform in a Transformation Era' in FTS19980506000321, Beijing Jiefangjun Bao (in Chinese), 21 April 1998.

Barrass, Gordon S. *The Great Cold War*, Stanford, CA: Stanford University Press, 2009.

Barth, Boris, Dolchstosslegenden und Politische Desintegration: Das Trauma der deutschen Niederlage im Ersten Weltkrieg 1914–1933, Schriften Des Bundesarchivs, Düsseldorf: Droste, 2003.

Bartov, Omer, 'From *Blitzkrieg* to Total War: Controversial Links between Image and Reality' in Ian Kershaw and Moshe Lewin (eds) *Stalinism and Nazism: Dictatorships in Comparison*, Cambridge: Cambridge University Press, 1997, pp. 158–84.

Baylis, J., J. Wirtz, E. Cohen and C. Gray, (eds), *Strategy in the Contemporary World: Introduction to Strategic Studies*, Oxford/New York: Oxford University Press, 2nd edn., 2007.

Beckett, Ian F. W. and John Pimlott (eds), *Armed Forces and Modern Counter-insurgency*, New York: St. Martin's, 1985.

Bengio, Ofra, *Saddam's Iraq – political discourse and the language of power* (in Hebrew), Tel Aviv University Press, 1999,

Biddle, Stephen, 'The Past as Prologue: Assessing Theories of Future Warfare' in *Security Studies* 8:1, 1998, pp. 1–74.

Biddle, Stephen, *Military Power: Explaining Victory and Defeat in Modern Battle*, Princeton, NJ: Princeton University Press, 2004.

Biddle, Stephen, 'Iraq, Afghanistan, and American Military Transformation' in John Baylis, James Wirtz, Colin S. Gray and Eliot Cohen, *Strategy in the Contemporary World*, Oxford: Oxford UP, 2007

Biddle, Stephen, 'The New U.S. Army/Marine Corps Counterinsurgency Field Manual as Political Science and political Praxis' in *Perspectives on Politics*, 6:2, June 2008.

Bishop, Patrick, *3 Para: Afghanistan, Summer 2006*, London: HarperPress, 2007.

Bjerga, Kjell Inge, *Unity as a Weapon: The Commander in Chief North Norway 1948–2002* (in Norwegian), Bergen: Eide, 2002.

Bjerga, Kjell Inge and Gullow Gjeseth, 'The Home Guard and the Army: a Divided or Unified Territorial Defence?' (in Norwegian) in *Oslo Files on Defence and Security*, 2, 2010.

Bjerga, Kjell Inge and Kjetil Skogrand, 'Securing small-state interests: Norway in NATO' in V. Mastny, Sven G. Holtsmark and A. Wenger (eds), *War plans and alliances in the Cold War*, London: Routledge, 2006, pp. 218–40.

Bland, Douglas L., 'Canada's Officer Corps: New Times, New Ideas', CDA Institute XVth Annual Seminar, 1999.

Blatt, Joel, *The French Defeat of 1940: Reassessments*. Providence, R.I.: Berghahn Books, 1998.

Bolin, Anna, *Political–Military Relations, An introduction to a field of study*, Stockholm: Försvarshögskolan, 2004.

Boot, Max, *War Made New: Technology, Warfare, and the Course of History, 1500 to Today*, New York: Gotham Books, 2006.

Booth, Ken, *Strategy and Ethnocentrism*, New York: Holmes & Meier, 1979.

Bracher, Karl Dietrich, *Die Auflösung der Weimarer Republik: Eine Studie zum Problem des Machtverfalls in der Demokratie*, Institut Für Politische Wissenschaft, Stuttgart: Ring-Verlag, 1955.

Brodie, Bernard, *Strategy in the Missile Age*, Princeton: Princeton University, 1965.

Brodie, Bernard, *War and Politics*, New York: Macmillan, 1973.

Brom, Shlomo and Meir Elran (eds), *The Second Lebanon War: Strategic Perspectives*, Tel Aviv: The Institute for National Security Studies, 2007.

Bronfeld, Saul, 'Fighting Outnumbered: The Impact of the Yom Kippur War on the U.S. Army' in *Journal of Military History*, 71:2, April 2007: pp. 465–98.

Bryant, William D., *The Iranian way of war*, Alabama: Air Force University, 2007.

Bucholz, Arden, *Hans Delbrück & the German Military Establishment: War Images in Conflict*, Iowa City: University of Iowa Prcss, 1985.
Builder, Carl H., *The Masks of War: American Military Styles in Strategy and Analysis*, Baltimore: Johns Hopkins University Press, 1989.
Butler, Robert, Charles Hawkins, and Timothy Thomas, 'West Meets East: Chinese and Western Researchers Exchange Views on the Revolution in Military Affairs' in *Historical Evaluation and Research Organization*, Library, 1999.
Cao Qisheng and Huang Chao, 'Jinan Theatre Carries Out Special-Topic Study on Frequency Use' (in English) in *PLA Daily*, 12 June 2009.
Citino, Robert Michael, *Blitzkrieg to Desert Storm: The Evolution of Operational Warfare*, Modern War Studies, Lawrence, Kan.: University Press of Kansas, 2004.
Citino, Robert Michael, *The Path to Blitzkrieg: Doctrine and Training in the German Army, 1920–1939*, Boulder, Colo.: Lynne Rienner Publishers, 1999.
Clark, Asa A, *The Defense Reform Debate: Issues and Analysis*, Baltimore: Johns Hopkins University Press, 1984.
Clausewitz, Carl von, *On War*, ed. and trans. by Michael Howard and Peter Paret, Princeton: University Press, 1976.
Coghlan, Tom, 'The Taliban in Helmand: An Oral History' in Antonio Giustozzi (ed.), *Decoding the New Taliban*, London: Hurst, 2009.
Cohen, Eliot A., 'Come the Revolution' in *National Review*, 47:14, July 1995.
Cohen, Eliot A., 'A Revolution in Warfare' in *Foreign Affairs*, 75:2, 1996, pp. 37–54.
Cohen, Eliot A., 'Change and Transformation in Military Affairs' in *Journal of Strategic Studies*, 27:3, 2004.
Cohen, Eliot A. and John Gooch, *Military Misfortunes: The Anatomy of Failure in War*, New York: Free Press, 1990.
Cooley, Alexander, *Logics of Hierarchy: The Organization of Empires, States, and Military Occupations*, Cornell: University Press, 2005.
Cooper, Matthew, *The German Army, 1933–1945: its Political and Military Failure*, London: Macdonald and Jane's, 1978.
Cordesman, Antony H. and K. R. Al-Rodhan, 'The Gulf military forces in the era of asymmetric warfare – Iran', Center for Strategic and International Studies, working draft for review and comments, Washington, DC, 2006.
Corum, James S. *The Roots of Blitzkrieg: Hans Von Seeckt and the German Military Reform*, Lawrence, Kan.: University Press of Kansas, 1992.
Craig, Gordon Alexander, *The Politics of the Prussian Army 1640–1945*, Oxford: Clarendon Press, 1955.
Creveld, Martin van, *The Transformation of War*, New York: Free Press, 1991.
Czege de, Huba Wass and Antulio J. Echevarria II, 'Toward a Strategy of Positive Ends' in *National Security Studies Quarterly*, 8:1, Winter 2002.
Dai, Qingmin, 'On Seizing Information Supremacy', in CPP20030728000209, Beijing Zhongguo Junshi Kexue (in Chinese), 20 April 2003.
Dai, Qingmen, 'Discourse on Armed Forces Informationization Building and Information Warfare Building' in *On the Chinese Revolution in Military Affairs*, ed. Shen Weiguang, Beijing: New China Press, 2004.
Dawkins, Richard, *The God Delusion*, London: Bantam Press, 2006.
Deist, Wilhelm, 'The Road to Ideological War: Germany, 1918–1945' in MacGregor Knox, Williamson Murray and Alvin H. Bernstein (eds), *The Making of Strategy: Rulers, States, and War*, Cambridge: Cambridge University Press, 1994, pp. 352–92.

Deist, Wilhelm, '"*Blitzkrieg*" Or Total War? War Preparations in Nazi Germany' in Roger Chickering and Stig Förster (eds), *The Shadows of Total War: Europe, East Asia, and the United States, 1919–1939*, Cambridge: University Press, 2003, pp. 271–84.

Denrell, Jerker and James G. March, 'Adaptation as Information Restriction: The Hot Stove Effect' in James G. March (ed.), *Explorations in Organizations*, Stanford: University Press, 2008.

Desportes, Vincent, *La Guerre probable: Penser Autrement*, Collection Stratégies & Doctrines, Paris: Economica, 2007.

DiNardo, R. L. 'German Armour Doctrine: Correcting the Myths' in *War in History*, 3:4, 1996, pp. 384–97.

Ding, Haiming and Sun Zhaoqiu, 'First Appearance of "Informatized Blue Army" in the Training Field' in CPP20050711000095 Beijing Jiefangjun Bao (Internet version, in Chinese), 11 July 2005.

Doubler, Michael D., *Closing with the Enemy: How GIs Fought the War in Europe, 1944–1945*, Kansas: University of Kansas Press, 1994.

Downey, J. and G. Murdock, 'The Counter Revolution in military affairs: the globalization of guerrilla warfare' in Daya K. Thussu and Des Freedman (eds), *War and the media: Reporting Conflict 24/7*, London: Sage Publications, 2003,

Dupuy, Trevor N., *A Genius for War: The German Army and General Staff, 1807–1945*, Englewood Cliffs, NJ: Prentice-Hall, 1977.

Eagleton, Terry, *The Idea of Culture*, Oxford: Blackwell, 2000.

Eden, Lynn, *Whole World on Fire: Organizations, Knowledge, and Nuclear Weapons Devastation*, Cornell University Press, 2004.

Edgerton, David, *The Shock of the Old: Technology and Global History since 1900*, Profile Books, 2006.

Egeland, Jan, *Impotent superpower – potent small state: potentials and limitations of human rights objectives in the foreign policies of the United States and Norway*, Oslo: University Press, 1988.

Eisenstadt, Michael and Kenneth M. Pollack, 'Armies of snow and armies of sand: the impact of Soviet military doctrine on Arab militaries' in *The Middle East Journal*, 55:4, pp. 549–70.

English, Allan D., *Understanding Military Culture, A Canadian Perspective*, Montreal & Kingston: McGill-Queen's University Press, 2004.

English, Allan D., Daniel Gosselin, Howard Coombs and Laurence M. Hickey (eds), *The Operational Art: Canadian Perspectives, Context and Concepts*, Kingston, Ontario: Canadian Defence Academy, 2005.

Erickson, Andrew S. and David D. Yang, 'On the Verge of a Game-Changer' in *Proceedings*, 135:5, May 2009.

Exum, Andrew, 'Hezbollah at War: A Military Assessment' in *Policy Focus*, 63, The Washington Institute for Near East Policy, December 2006.

Farrell, Theo, 'Humanitarian Intervention and Peace Operations' in J. Baylis, J. Wirtz, E. Cohen and Colin S. Gray, *Strategy in the Contemporary World*, 2nd edn., Oxford: University Press, 2007.

Farrell, Theo and Stuart Gordon, 'COIN Machine: the British Military in Afghanistan' in *Orbis*, 53:4, 2009.

Farrell, Theo and Terry Terriff (eds), *The Sources of Military Change, Culture, Politics, Technology*, Boulder: Lynne Rienner, 2002.

Ferguson, Niall, *The Cash Nexus: Money and Power in the Modern World, 1700–2000*, New York: Basic Books, 2001.

Fergusson, James, *A Million Bullets*, Bantham Press, 2008.
Fisher, Richard, *China's Military Modernization: Building for Regional and Global Reach*, Westport CT: Praeger Security International, 2008.
Fitzgerald, Mary C. 'Marshall Ogarkov on the Modern Theater Operation' in *Conflict Quarterly*, Summer 1986.
Fitzsimmonds, James R. and Jan M. Van Tol, 'Revolutions in Military Affairs' in *Joint Force Quarterly*, Spring 1994, pp. 24–31.
Foch, Ferdinand, *The Principles of War*, London: Chapman & Hall, 1920.
Förster, Jürgen, 'Wehrmacht, Krieg Und Holocaust', in Müller and Volkmann, *Wehrmacht*, pp. 948–63.
Freedman, Lawrence and Efraim Karsh, *The Gulf Conflict, 1990–1991: Diplomacy and War in the New World Order*, Princeton: University Press, 1991.
Frieser, Karl-Heinz, *Blitzkrieg-Legende: Der Westfeldzug 1940*, 2. Aufl. Operationen Des Zweiten Weltkrieges, Munich: R. Oldenbourg, 1996.
Frieser, Karl-Heinz, 'Die Deutschen Blitzkriege: Operativer Triumph – Strategische Tragödie' in Rolf-Dieter Müller and Hans-Erich Volkmann (eds), *Die Wehrmacht: Mythos und Realität*, Munich: Oldenbourg, 1999, pp.182–96.
Fukuyama, Francis, *The End of History and the Last Man*, New York: Free Press, 1992.
Fuller, J. F. C., *On Future Warfare*, London: Praed, 1928.
Fuller, J. F. C., *Towards Armageddon*, London: Dickson, 1937.
Fuller, J. F. C., *Armament and History*, London: Eyre, 1946.
Fuller, J. F. C., *A Military History of the Western World*, Vol. 3, New York: DaCapo, 1956.
Garraud, Philippe, 'L'idéologie de la "défensive" et ses effets stratégiques: le rôle de la dimension cognitive dans la défaite de 1940' in *Revue française de science politique*, 54:5, 2004, pp. 781–810.
Garver, Eugene, *Aristotle's Rhetoric, An Art of Character*, Chicago: University of Chicago Press, 1994.
Gat, Azar, *British Armour Theory and the Rise of the Panzer Arm: revising the Revisionists*, St Anthony's Series, London: Macmillan Press, 2000.
Gat, Azar, 'Ideology, National Policy, Technology and Strategic Doctrine between the World Wars' in *Journal of Strategic Studies*, 24:3, 2001, pp. 1–18.
Gat, Azar, *A History of Military Thought: from the Enlightenment to the Cold War*, Oxford: Oxford University Press, 2001.
Gentile, Gian P., 'Learning, Adapting and the Perils of the New Counter-insurgency' in *Survival*, 5:6, 2010.
Geyer, Michael, 'German Strategy in the Age of Machine Warfare, 1914–1945' in Peter Paret, Gordon A. Craig and Felix Gilbert (eds), *Makers of Modern Strategy: from Machiavelli to the Nuclear Age*, Princeton, N.J.: Princeton University Press, 1986, pp. 527–97.
Gilboa, Amos, 'The Arab security strategies following the Yom Kippur War' (in Hebrew) in *Maarchot*, 361, November 1998.
Gjeseth, Gullow, *Transforming the Norwegian Army 1990–2005* (in Norwegian), Bergen: Vigmostad og Bjørke, 2008.
Goldman, Emily O., 'The Spread of Western Military Models to Ottoman Turkey and Maiji Japan' in Theo Farrell and Terry Terriff (eds), *The Sources of Military Change. Culture, Politics, Technology*, Boulder, CO: Lynne Rienner, 2002.
Goldman, Emily O., 'International Competition and Military Effectiveness: Naval Air Power, 1919–1945' in Risa Brooks and Elizabeth A. Stanley (eds), *Creating*

Military Power: The Sources of Military Effectiveness, Standford, CA: Stanford University Press, 2007.

Gompert, David C., Richard L. Kugler and Martin C. Libiki, *Mind the Gap: Promoting a Transatlantic Revolution in Military Affairs*, Washington, DC: National Defense University, 1999.

Gooch, John, 'Military Doctrine and Military History' in John Gooch (ed.), *The Origins of Contemporary Doctrine, Strategic and Combat Studies Institute*, 30, September 1997.

Gordon, Michael R. and Bernard E. Trainor, *The Generals' War*, Boston: Little Brown, 1995.

Grant, Rebecca, *Airpower and the Total Force: The Gift of Time*, Arlington, VA: IRIS Independent Research, 1998.

Gray, Chris Hables, *Postmodern War – The New Politics of Conflict*, New York: The Guilford Press, 1997.

Gray, Colin S. *Strategy for Chaos: Revolutions in Military Affairs and the Evidence of History*, London: Frank Cass, 2002.

Gray, Colin S. *Another Bloody Century: Future Warfare*, London: Weidenfeld and Nicolson, 2005.

Grey, Stephen, *Operation Snakebite*, Viking, 2009.

Griffiths, Paddy, *Battle Tactics on the Western Front: The British Army's Art of Attack, 1916–18*, Yale NH: Yale University Press, 1994.

Grissom, Adam, 2006. 'The Future of Military Innovation Studies' in *Journal of Strategic Studies*, 29:5, 2006.

Gross, Gerhard P., 'Das Dogma der Beweglichkeit. Überlegungen zur Genese der deutschen Heerestaktik im Zeitalter der Weltkriege' in Bruno Thoss and Hans Erich Volkmann (eds.) *Erster Weltkrieg, Zweiter Weltkrieg: Ein Vergleich: Krieg, Kriegserlebnis, Kriegserfahrung in Deutschland*, Paderborn: F. Schöningh, 2002, pp. 143–66.

Guangqian, Peng and Yao Youzhi (eds), *The Science of Military Strategy*, English Language Edition, Beijing: Military Science Publishing House, 2005.

Gunsburg, Jeffrey A., 'Coupable ou non? Le rôle du général Gamelin dans la défaite de 1940' in *Revue historique des armées*, 4, 1979, pp. 145–63.

Gurr, Nadine and Benjamin Cole, *The New Face of Terrorism: Threats from Weapons of Mass Destruction*, London: Tauris, 2000.

Haaland, Torunn Laugen, *Small Forces with a Global Outreach. Role Perceptions in the Norwegian Armed Forces after the Cold War*, Oslo: Unipub, 2008.

Haas, Marcel de., 'Dutch Defence Doctrine' in *The Officer Magazine*, 18:1, 2006.

Habeck, Mary R., *Storm of Steel: the Development of Armor Doctrine in Germany and the Soviet Union, 1919–1939*, Cornell Studies in Security Affairs, Ithaca, NY: Cornell University Press, 2003.

Halberstam, David, *War in a Time of Peace: Bush, Clinton, and the Generals*, Scribner, 2001.

Hallion, Richard P., *Storm Over Iraq: Air Power and the Gulf War*, Washington, DC: Smithsonian Institution Press, 1992.

Harari, Yuval Noah, *The Ultimate Experience, Battlefield Revelations and the Making of Modern War Culture, 1450–2000*, New York: Palgrave Macmillan, 2008.

Harkabi, Yehoshafat, *Arab strategies and Israel's response*, New York: The Free Press, 1977.

Harris, J.P., 'The Myth of Blitzkrieg', *War in History*, 2, 1995, pp. 335–52.

Hartmann, Christian, 'Wehrmacht im Ostkrieg: Front und Militärisches Hinterland 1941/42' in *Quellen und Darstellungen zur Zeitgeschichte*, 75, Munich: Oldenbourg, 2009.

Hashim, Ahmed, 'Iraq's Civil War' in *Current History – a Journal of Contemporary World Affairs*, January 2007.

Hawkins, Charles F., 'The People's Liberation Army Looks to the Future' in *Joint Forces Quarterly*, 25, Summer 2000.

Headrick, Daniel, *The Tools of Empire: Technology and European Imperialism in the Nineteenth Century*, New York: Oxford UP, 1981.

Heer, Hannes and Klaus Naumann (eds), *Vernichtungskrieg: Verbrechen der Wehrmacht 1941–1944*, Hamburg: Hamburger Edition, 1995.

Herbert, Ulrich, *Hitler's Foreign Workers: Enforced Foreign Labor in Germany under the Third Reich*, Cambridge: Cambridge University Press, 1997.

Hezhen, Xu, 'Focus on Psychological War Under the Background of Larger Military Strategy', in CPP20001211000122, Beijing Zhongguo Junshi Kexue (in Chinese), 20 October 2000.

Hilburn, Matt, 'Policing the Insurgents' in *Seapower Magazine*, March 2006.

Hobson, Rolf, *War and Strategic Thought in Europe 1500–1945: Civil Society, State system and Military Theory*, (in Norwegian), Oslo: Cappelen akademisk, 2005.

Hoffman, Frank, 'Hybrid warfare and challenges' in *Joint Forces Quarterly*, 51:1, 2009.

Hofmann, George F. and Donn A. Starry, *Camp Colt to Desert Storm: The History of U.S. Armored Forces*, Kentucky: University Press, 1999.

Holbeche, Linda, *Understanding Change, Theory, Implementation and Success*, Oxford: Elsevier, 2006.

Holden-Reid, Brian, 'Warfighting Doctrine and the British Army', Army Code 71565, Army Doctrine Publication, Vol. 1, Operations, HQDT/18/34/46, London: MoD, 1A-1–1A-15, 1994.

Honderich, Ted, *The Oxford Companion to Philosophy*, Oxford: Oxford University Press, 1995.

Hooker, Richard D., *Maneuver Warfare: an Anthology*, Novato, CA: Presidio Press, 1993.

House of Commons Defence Committee, *U.K. Operations in Afghanistan*, Thirteenth Report of Session 2006–2007, HC 408, London: TSO, 2007.

Hughes, Daniel J., *Moltke On the Art of War, Selected Writings*, New York: Ballantine Books, 1993.

Hull, Isabel V., *Absolute Destruction: Military Culture and Practices of War in Imperial Germany*, Cornell University Press, 2005.

Hundley, Richard O., *Past Revolutions, Future Transformations: what Can the History of Revolutions in Military Affairs Tell Us about Transforming the U.S. Military?* Santa Monica, CA: Rand, 1999.

Huntington, Samuel P., *The Soldier and the State; the Theory and Politics of Civil–Military Relations*, Cambridge, Mass.: Belknap Press of Harvard University Press, 1957.

Hürter, Johannes, 'Hitlers Heerführer: Die deutschen Oberbefehlshaber im Krieg gegen die Sowjetunion 1941/42' in *Quellen und Darstellungen zur Zeitgeschichte*, Munich: Oldenbourg, 2006.

Ignatieff, Michael, *Virtual War: Kosovo and Beyond*, New York: Metropolitan, 2000.

Ikle, Fred and Albert Wohlstetter, *Discriminate Deterrence*, Washington DC: Government Printing Office, 1988,

Jersak, Tobias. 'Blitzkrieg Revisited: A New Look at Nazi War and Extermination Planning' in *The Historical Journal*, 43, 2000, pp. 565–82.

Johnson, David E., *Learning Large Lessons: The Evolving Roles of Ground Power and Air Power in the Post-Cold War World*, Santa Monica, CA: RAND, 2007.

Johnston, Paul, 'Doctrine Is Not Enough: The Effect of Doctrine on the Behaviour of Armies' in *Parameters*, Autumn 2000.

Kahl, Colin H., 'In the Crossfire or the Crosshairs? Norms, Civilian Casualties, and U.S. Conduct in Iraq' in *International Security*, 32:1, 2007.

Kahn, Herman, *On Thermonuclear War*, Princeton: Princeton University, 1960.

Kahn, Herman, *On Escalation*, New York: Praeger, 1965.

Kaldor, Mary, *New and Old Wars: Organized Violence in a Global Era*, Stanford: University Press, 2007.

Kaplan, Fred M. *The Wizards of Armageddon*, New York: Simon and Schuster, 1983.

Keaney, Thomas and Eliot A. Cohen, *Gulf War Air Power Survey Summary Report*, Washington DC: U.S. Government Printing Office, 1993.

Khalidi, A.S. and H. Agha, 'The Syrian doctrine of strategic parity' in H. H. Sanders, *The Middle East in global perspective*, Colorado: Westview Press, 1991.

Kier, Elisabeth, *Imagining War. French and British Military Doctrines Between the Wars*, Jersey: Princeton University Press, 1997.

Kiley, Sam, *Desperate Glory: At War in Helmand with Britain's 16 Air Assault Brigade*, London: Bloomsbury, 2009.

Kiszely, John, 'Learning about Counterinsurgency' in *RUSI Journal*, 151:6, 2006, pp. 16–21.

Knox, MacGregor, Williamson Murray, and Alvin H. Bernstein (eds), *The Making of Strategy: Rulers, States, and War*, Cambridge: Cambridge University Press, 1994.

Kober, Aver, 'The Israeli Defense Forces in the Second Lebanon War: Why the Poor Performance?' in *Journal of Strategic Studies*, 31:1, 2008.

Koenen, Gerd, *Der Russland-Komplex: Die Deutschen und der Osten 1900–1945*, Munich: Beck, 2005.

Kokoshin, Andrei A., *O revoliutcii v voennom dele v proshlom i nastoiaschem*, Moscow: URSS, 2006.

Kokoshin, Andrei A., *Innovatcionnye vooruzhennye sily i revoliutciia v voennom dele*, Moscow: URSS, 2009.

Krause, Michael D. and R. Cody Phillips (eds), *Historical Perspectives of the Operational Art*, Washington, DC: Center of Military History, 2005.

Krepinevich, Andrew F., 'Keeping Pace with the Military Technical Revolution' in *Issues in Science and Technology*, Summer 1994.

Krepinevich, Andrew F., 'Cavalry to Computer: The Pattern of Military Revolutions' in *The National Interest*, 37, Fall 1994, pp. 30–42.

Krepinevich, Andrew F., *The Military–Technological Revolution: A Preliminary Assessment*, Washington, DC: Center for Strategic and Budgetary Assessments, 2002.

Kroener, Bernhard R., 'Der "Erfrorene Blitzkrieg". Strategische Planungen Der Deutschen Führung Gegen Die Sowjetunion Und Die Ursachen Ihres Scheiterns' in Bernd Wegner (ed.), *Zwei Wege nach Moskau: Vom Hitler-Stalin-Pakt bis zum 'Unternehmen Barbarossa'*, Munich: Piper, 1991, pp. 133–48.

Kronvall, Olof Kronvall, 'Finally Eating Soup with a Knife? A Historical Perspective on the U.S. Army's 2006 Counterinsurgency Doctrine' in *Oslo Files on Defence and Security*, 5, 2007.

Kulick, Amir, 'The IDF's Combat Approach vs. Hezbollah' in *Strategic Assessment*, 9:3, The Institute for National Security Studies, November 2006.

Kun, Liu, 'China: Torpedo Stealth' in FTS19960809000630, Beijing, Jianchuan Zhishi (Naval and Merchant Ships – in Chinese), 9 August 1996.

Lacey, Jim, *The Canons of Jihad: Terrorists' Strategy for Defeating America*, Annapolis: Naval Institute Press, 2003.

Langlois, H., *Lessons From Two Recent Wars, The Russo-Turkish and South African Wars*, translated for the General Staff, War Office, London: HMSO, 1909.

Laqueur, Walter, *Guerilla Warfare: A Historical and Critical Study*, Boulder: Westview, 1984.

Lederman, Gordon Nathaniel, *Reorganizing the Joint Chiefs of Staff: The Goldwater–Nichols Act of 1986*, Westport CT: Greenwood Press, 1999.

Leonhard, Robert R., *The Art of Maneuver: Maneuver-Warfare Theory and Airland Battle*, Novato, CA: Presidio Press, 1991.

Leonhard, Robert L. (Foreword by Colonel J. R. McDonough), *Fighting by Minutes. Time and Art of War*, Westport Connecticut: Praeger, 1994.

Levitt, Barbara and James G. March, 'Chester I. Barnard and the Intelligence of Learning' in Oliver Williamson (ed.), *Organization Theory: From Chester Barnard to the Present and Beyond*, Oxford: University Press, 1995.

Levy, Jack, *War in the Modern Great Power System, 1495–1975*, Lexington: UP of Kentucky, 1983.

Lewis, Bernard, *The Crisis of Islam: Holy War and Unholy Terror*, London: Weidenfeld and Nicolson, 2003.

Lewy, Guenter, *America in Vietnam*, New York: Oxford University Press, 1978.

Libicki, Martin and Stuart Johnson (eds), *Dominant Battlespace Knowledge: The Winning Edge*, Washington DC: National Defense University, 1996.

Liddell Hart, Basil, *Strategy*, New York, N.Y.: Meridian, 2nd revised edn., 1991.

Liddell Hart, Basil, 'Strategy: the indirect approach' in Thomas G. Mahnken and Joseph Maiolo (eds), *Strategic Studies – a Reader*, Oxon/New York: Routledge, 2008.

Lind, William S., *Maneuver Warfare Handbook*, Westview Special Studies in Military Affairs, Boulder, Colo.: Westview Press, 1985.

Liping, Xi, 'China, U.S., Japan Strategic Relations: Striving for Win-win and Avoiding Security Dilemmas' in CPP20071109587001, Beijing Shijie Jingji Yu Zhengzhi, (in Chinese), September 2007.

Locher, James R., *Victory on the Potomac: The Goldwater-Nichols Act Unifies the Pentagon*, College Station, Texas: A&M University Press, 2002.

Lock-Pullan, Richard, 'How to rethink war: conceptual innovation and AirLand Battle Doctrine' in *Journal for Strategic Studies*, 28:4, August 2005, pp. 679–702.

Løvold, Andras, Small state challenges in international relations (in Norwegian), in Internasjonal Politikk, 1, 2004.

Luard, Evan, *War in International Society*, London: Tauris, 1986.

Luttwak, Edward N., 'The Operational Level of War' in *International Security*, 5, 1981: pp. 61–79.

Luvaas, Jay, 'Some Vagrant Thoughts on Doctrine' in *Military Review*, March 1986.

Mackay, Andrew and Steve Tatham, *Behavioural Conflict: From General to Strategic Corporal: Complexity, Adaptation and Influence*, Shrivenham Paper No. 9, 2009.

Mäder, Markus, *In Pursuit of Conceptual Excellence, The Evolution of British Military–Strategic Doctrine in the Post-Cold War Era, 1989–2002*, Bern: Studies in Contemporary History and Security Policy Vol. 13, 2004.

Makovsky, David and Jeffrey White, 'Lessons and Implications of the Israel–Hezbollah War: A Preliminary Assessment' in *Policy Focus*, 60, The Washington Institute for Near East Policy, October 2006.

Malik, S. K., *The Qur'anic Concept of War*, Dehli: Adam Publisher, 1992.

Malkasian, Carter, 'Did the Coalition Need More Forces in Iraq? Evidence from Al Anbar' in *Joint Forces Quarterly*, 3rd Quarter, 2007.

March, James G. (ed.), *The Pursuit of Organizational Intelligence*, Blackwell, 1999.

Marshall, Andrew W., 'Forward' in *The Military-Technological Revolution: A Preliminary Assessment, 1991*, Center for Strategic and Budgetary Assessments, 2002.

Martin, Mike, 'The Importance of Cultural Understanding to the Military' in *British Army Review*, 147, 2009.

Mason, Timothy W., 'Internal Crisis and War of Aggression, 1938–1939' in Jane Caplan (ed.), *Nazism, Fascism and the Working Class*, Cambridge: Cambridge University Press, 1995, pp. 104–30.

May, Ernest R., *Strange Victory: Hitler's Conquest of France*, New York: Hill and Wang, 2000.

McCone, David R., Wilbur J. Scott and George R. Mastroianni, 'The 3rd ACR in Tal'Afar: Challenges and Adaptations' in *Strategic Studies Institute*, Carlisle PA: U.S. Army War College, January 2008.

McInnes, Colin, *Hot War, Cold War: The British Army's Way of Warfare, 1945–1995*, London: Brassey's, 1996.

McInnes, Colin and John Stone, 'The British Army and Military Doctrine' in Michael Duffy, Theo Farrell and Geoffrey Sloan (eds), *Doctrine and Military Effectiveness*, Exeter: SPSG, 1997.

McIvor, Anthony D. (ed.), *Rethinking the Principles of War*, Annapolis: Naval Institute Press, 2005.

McKitrick, Jeffrey S. and Peter W. Chiarelli, 'Defense reform: An Appraisal' in Clark, Chiarelli, McKitrick and J.W. Reed (eds), *The Defense Reform Debate, Issues and Analysis*, Baltimore: The Johns Hopkins University Press, 1984.

McMaster, H. R., 'On War: Lessons to Be Learned' in *Survival*, 50, 2008, pp. 19–30.

Miksche, F. O., *Blitzkrieg*, London: Faber & Faber, 1941.

Milward, Alan S., *The German Economy at War*, Athlone Press: London, 1965.

Mishal, Shaul and Avraham Sela, *The Palestinian Hamas*, Columbia: University Press, 2000.

Mitchell, B. R., *International Historical Statistics, Europe 1750–1988*, New York: Stockton, 1992.

Murray, Williamson, 'German Army Doctrine, 1918–1939, and the Post-1945 Theory of "Blitzkrieg Strategy"' in Carole Fink, Isabel V. Hull and MacGregor Knox (eds), *German Nationalism and the European Response, 1890–1945*, Norman: University of Oklahoma Press, 1985, pp. 71–94.

Murray, Williamson, 'Armoured Warfare: The British, French and German Experiences' in Williamson Murray and Allan R. Millett (eds) *Military Innovation in the Interwar Period*, Cambridge: Cambridge University Press, 1996, pp. 6–49.

Murray, Williamson, 'Thinking About Revolutions in Military Affairs' in *Joint Force Quarterly*, Summer 1997, pp. 69–76.

Murray, Williamson, 'May 1940: Contingency and Fragility of the German RMA' in MacGregor Knox and Murray Williamson (eds), *The Dynamics of Military Revolution, 1300–2050*, Cambridge: Cambridge University Press, 2001, pp. 154–74.

Murray, Williamson and Allan R. Millet (eds), *Military Innovation in the Interwar Period*, Cambridge University Press, 1996.

Nagl, John A, *Learning to Eat Soup with a Knife: Counterinsurgency Lessons from Malaya and Vietnam*, Chicago IL: Chicago University Press, 2002.

Nation, Craig R., *War in the Balkans, 1991–2002*, Carlisle Barracks: U.S. Army War College, 2003, p. ix.

Naveh, Shimon, *In Pursuit of Military Excellence: The Evolution of Operational Theory*, London: Frank Cass, 1997.

Neugebauer, Karl-Volker, 'Operatives Denken zwischen dem Ersten und Zweiten Weltkrieg' in *Operatives Denken und Handeln in deutschen Streitkräften im 19. und 20. Jahrhundert*, Herford, Bonn: E. S. Mittler & Sohn, 1988, pp. 97–122.

Newmann, Ronald E, *The Other War*, Washington DC: Potomac Books, 2009.

Newmyer, Jacqueline A., 'Oil, Arms, and Influence: The Indirect Strategy Behind Chinese Military Modernization' in *Orbis*, Spring 2009.

Nielsen, Suzanne C. and Don M. Snider (eds.), *American Civil–Military Relations: The Soldier and the State in a New Era*, Baltimore: Johns Hopkins University Press, 2009.

Nissestad, Odd Arne, *Leadership development: an empirical study of effectiveness of the leadership development program at The Royal Norwegian Naval Academy and its impact on preparing officers to execute leadership in today's conflicts and the conflicts in the years ahead*, Bergen: NHH, 2007.

Nolan, Victoria L, *Command and Culture: Military leadership and the Evolution of the British Army's Approach to Small Wars*, Ph.D. thesis, King's College London, 2009.

Norton, Augustus Richard, *Hezbollah: A Short History*, Princeton: University Press, 2007.

Olsen, John Andreas, *Strategic Air Power in Desert Storm*, London: Frank Cass, 2003.

Olsen, John Andreas, *John Warden and the Renaissance of American Air Power*, Washington, DC: Potomac, 2007.

Osanka, Franklin Mark (ed.), *Modern Guerrilla Warfare: Fighting Communist Guerrilla Movements, 1941–1961*, Glencoe, Il: Free Press, 1962.

Osgood, Robert E., *Limited War: The Challenge to American Strategy*, Chicago: University of Chicago, 1957.

Osgood, Robert E., *Limited War Revisited*, Boulder: Westview, 1979.

Osinga, Frans, 'The Rise of Military Transformation' in Terry Terriff, Frans Osinga and Theo Farrell (eds), *A Transformation Gap? American Innovations and European Military Change*, Stanford CA: Stanford University Press, 2010.

Overy, R. J., *War and Economy in the Third Reich*, Oxford: Clarendon Press, 1994.

Owens, Bill, *Lifting the Fog of War,* New York: Farrar, Strauss, and Giroux, 2000.

Palazzo, Albert, *From Moltke to Bin Laden, The Relevance of Doctrine in the Contemporary Military Environment*, Australian Government, Department of Defence, Land Warfare Studies Centre, 2008.

Palmer Harik, Judith, *Hezbollah: the Changing Face of Terrorism*, I. B. Tauris, 2005.

Pape, Robert A., 'The True Worth of Air Power' in *Foreign Affairs* 83:2, March/April 2004.

Paret, Peter, *Yorck and the Era of Prussian Reform 1807–1815*, Princeton University Press, 1966.

Parkinson, Roger, *Clausewitz – A Biography* (1971), New York: Cooper Square Press, 2002.

Pätzold, Kurt, *Ihr waret die besten Soldaten: Ursprung und Geschichte einer Legende*, Leipzig: Militzke, 2000.

Peng Guangqian and Yao Youzhi (eds), *The Science of Military Strategy*, English Language Edition, Beijing: Military Science Publishing House, 2005.

Peters, Ralph, 'Lessons from Lebanon: The New Model Terrorist Army' in *Armed Forces Journal*, 144, October 2006.

Pillsbury, Michael, *China Debates the Future Security Environment*, Washington DC: National Defense University Press, 2000.

Pohl, Dieter, 'Die Herrschaft der Wehrmacht: Deutsche Militärbesatzung und einheimische Bevölkerung in der Sowjetunion 1941–1944' in *Quellen und Darstellungen zur Zeitgeschichte*, 71, Munich: Oldenbourg, 2008.

Pollack, Kenneth M., *Arabs at war, military effectiveness 1948–1991*, Nebraska: University of Nebraska Press, 1991.

Pope-Hennessy, Ladislaus, 'The British Army and Modern Conceptions of War' in *The Edinburgh Review*, April 1911.

Pope-Hennessy, Ladislaus, 'The Place of Doctrine in War' in *The Edinburgh Review*, January 1912.

Porch, Douglas, 'Military "Culture" And the Fall of France in 1940' in *International Security*, 24, 2000, pp. 157–80.

Posen, Barry R., *The Sources of Military Doctrine: France, Britain, and Germany Between the World Wars*, Ithaca NY: Cornell University Press, 1984.

Posen, Barry R., 'Command of the Commons: The Military Foundation of U.S. Hegemony' in *International Security*, 28:1, Summer 2003.

PRC State Council, *China's National Defence in 2004*, Beijing: State Council Information Office, 2004.

PRC State Council, *China's National Defence in* 2006, Beijing: State Council Information Office, 2006.

Qassem, Naim, *Hizbullah – the story from within*, Saqi Books, 2005.

Raudzens, George, 'Blitzkrieg Ambiguities: Doubtful Usage of a Famous Word' in *War and Society*, 7, 1989, pp. 77–94.

Rayment, Sean, *Into the Killing Zone*, London: Constable, 2008.

Reid, Brian Holden, *A Doctrinal Perspective 1988–98*, The Occasional Paper No. 33, The Strategic and Combat Studies Institute, 1998.

Resende-Santos, Joao, *Neorealism, States and the Modern Mass Army*, Cambridge University Press, 2007.

Reynolds, Charles, 'Carl von Clausewitz and strategic theory' in *British Journal of International Studies*, 4, 1978.

Riste, Olav, *Norway's Foreign Relations*, Oslo: Universitetsforlaget, 2005.

Ritter, Gerhard, *Staatskunst und Kriegshandwerk; Das Problem des "Militarismus" in Deutschland*, München: R. Oldenbourg, 1954.

Rogers, Clifford J., ' "Military Revolutions" And "Revolutions in Military Affairs" ', Thierry Gongora and Harald Von Riekhoff (eds) *Toward a Revolution in Military Affairs?: Defense and Security at the Dawn of the Twenty-First Century*, Westport, Conn.: Greenwood Press, 2000, pp. 21–35.

Römer, Felix, *Der Kommissarbefehl: Wehrmacht und NS-Verbrechen an der Ostfront 1941/42*, Paderborn: Schöningh, 2008.

Romjue, John L., *American Army Doctrine for the Post-Cold War*, Fort Monroe: Military History Office, 1997.

Rosen, Stephen Peter, *Winning the Next War: Innovation and the Modern Military*, Cornell University Press, 1991.

Rotmann, Philipp, David Tohn and Jaron Wharton, 'Learning Under Fire: Progress and Dissent in the U.S. Military' in *Survival*, 51:4, 2009.
Saeveraas, Torgeir E. and Kjetil Henriksen, 'A Military Wonder Medicine? American "Maneuver Warfare" and Doctrine Development in Norway' (in Norwegian) in *Oslo Files on Defence and Security*, 1, 2007.
Scales, Robert H., *Certain Victory: The U.S. Army in the Gulf War*, Washington/London: Brassey's Inc., 1994.
Schelling, Thomas C., *Arms and Influence*, New Haven: Yale University, 1966.
Scott, John and Gordon Marshall, *Oxford Dictionary of Sociology*, Oxford: Oxford University Press, 2005.
Shen Weiguang, 'Trends in the Development of World Warfare – Reducing Destructive Force' in Shen Weiguang (ed.), *On the Chinese Revolution in Military Affairs*, Beijing: New China Press, 2004.
Showalter, Dennis, *Rifles and Railroads: Soldiers, Technology and the Unification of Germany*, Hamden, Conn.: Archon, 1975.
Showalter, Dennis, 'Ce que l'armée française avait compris de la guerre moderne' in Maurice Vaïsse (ed.), *Mai-juin 1940. Défaite française, victoire allemande, sous l'œuil des historiens étrangers*, Paris: Autrement, 2000, pp. 29–58.
Showalter, Dennis, 'More Than Nuts and Bolts: Technology and the German Army' in *The Historian*, 65, 2002, pp. 123–43.
Small, Melvin and David Singer, *Resort to Arms: International and Civil Wars, 1816–1980*, Beverly Hills: SAGE, 1982.
Smith, Niel, 'Retaking Sa'ad: Successful Counterinsurgency in Tal Afar' in *Armor Magazine*, July/August, 2007.
Smith, Rupert, *The Utility of Force, The Art of War in the Modern World*, London: Penguin, 2005.
Snyder, Jack, *The Ideology of the Offensive: Military Decision-Making and the Disaster of 1914*, Itacha: Cornell University Press, 1984.
Southby-Tailyour, Ewen, *Helmand, Afghanistan*, Ebury Press, 2008.
Spiller, Roger J., 'In the Shadow of the Dragon: Doctrine and the U.S. Army after Vietnam' in *RUSI Journal*, December 1997.
Stanley, Elizabeth A., *Evolutionary Technology in the Current Revolution in Military Affairs: The Army Tactical Command and Control System*, Carlisle, PA: Strategic Studies Institute, Army War College, 1998.
Starry, Donn A., 'A Tactical Evolution – FM 100–5', *Military Review*, 58, August 1978.
Starry, Donn A., 'The Legacy of Drummers, Warriors and Storytellers' in *Army Magazine*, 52, July 2002.
Stevenson, David, *Armament and the Coming of War: Europe, 1904–1914*, Oxford: Oxford UP, 1996.
Stout, Mark E., Jessica Huckabey and John R. Schindler, *The Terrorist Perspective Project*, Annapolis: Naval Institute Press, 2008.
Strachan, Hew, *European Armies and the Conduct of War*, London: Allen & Unwin, 1983.
Strachan, Hew, 'The Lost Meaning of Strategy' in *Survival*, 47, 2005, pp. 3–54.
Strachan, Hew (ed.), *Big Wars and Small Wars, The British Army and the Lessons of War in the 20th Century*, Abingdon: Routledge, 2006.
Strachan, Hew, 'Civil–Military Relations after Iraq' in *Survival*, 48, 2006, pp. 59–82.

Strachan, Hew, 'Norwegian Armed Forces Joint Operational Doctrine' in *Norsk Militært Tidsskrift*, 4, 2009.

Streit, Christian, *Keine Kameraden: Die Wehrmacht und die sowjetischen Kriegsgefangenen 1941–1945*, Studien Zur Zeitgeschichte Vol. 13, Stuttgart: Deutsche Verlags-Anstalt, 1978.

Sumida, Jon, *In Defence of Naval Supremacy: Finance, Technology and British Naval Policy*, Boston: Unwin, 1989.

Summers, Harry G. Jr., *On Strategy: A Critical Analysis of the Vietnam War*, Navato, Ca.: Presidio, 1995.

Taleb, Nassim Nicholas, *The Black Swan, The Impact of the Highly Improbable*, New York: Random House, 2007.

Terriff, Terry, 'Warriors and Innovators: Military Change and Organizational Culture in the U.S. Marine Corps', *Defence Studies* 6:2, 2006.

Tilford, Earl H. Jr., *The Revolution in Military Affairs: Prospects and Cautions*, Carlisle: Strategic Studies Institute, 1995.

Till, Geoffrey, *Seapower – A Guide for the Twenty-First Century*, London: Frank Cass, 2004.

Tootal, Stuart, *Danger Close: Commanding 3 PARA in Afghanistan*, London: John Murray, 2009.

Tooze, Adam, *The Wages of Destruction: The Making and Breaking of the Nazi Economy*, London: Allen Lane, 2006.

Trubowitz, Peter, Emily O. Goldman and Edward Rhodes (eds), *The Politics of Strategic Adjustment: Ideas, Institutions, and Interests*, Columbia University Press, 1999.

Tse-Tung, Mao, 'Problems of Strategy in China's revolutionary wars' (1938), cited in Mao Tse-Tung, *Selected military writings*, Foreign Languages Press, 1963.

Ucko, David, *The New Counterinsurgency Era: Transforming the U.S. Military for Modern Wars*, Washington DC: Georgetown University Press, 2009.

Valensi, Carmit, *The emergence of 'Hybrid actors'*, Tel Aviv: Tel Aviv University, 2011.

Vardi, Gil-li, 'The Enigma of German Operational Theory: The Evolution of Military Thought in Germany, 1919–1938', Ph.D. dissertation, The London School of Economics and Political Science, 2008.

Warden III, John A., *Air Campaign: Planning for Combat*, Washington, DC: National Defense University, 1989.

Watts, Barry D., *Six Decades of Guided Munitions and Battle Networks: Progress and Prospects*, Washington DC: CSBA, 2007.

Wegner, Bernd 'Erschriebene Siege. Franz Halder, Die "Historical Division" und die Rekonstruktion des Zweiten Weltkrieges im Geiste des Deutschen Generalstabes' in Ernst Willi Hansen, Gerhard Schreiber, Klaus-Jürgen Müller and Bernd Wegner (eds) *Politischer Wandel, Organisierte Gewalt und nationale Sicherheit: Beiträge zur neueren Geschichte Deutschlands und Frankreichs: Festschrift für Klaus-Jürgen Müller*, München: Oldenbourg, 1995, pp. 287–304.

Weigley, Russell F., *The American Way Of War: A History Of United States Military Strategy And Policy*, Bloomington, IN: Indiana University Press, 1973.

Wette, Wolfram, *Die Wehrmacht: Feindbilder, Vernichtungskrieg, Legenden*, Frankfurt am Main: S. Fischer, 2002.

Wheeler-Bennett, John, *The Nemesis of Power; the German Army in Politics, 1918–1945*, London: Macmillan, 1953.

Williamson, Oliver E (ed.), *Organization Theory: From Chester Barnard to the Present and Beyond*, Oxford: University Press, 1995.

Wilson, David and Gareth E Conway, 'The Tactical Conflict Assessment Framework' in *RUSI Journal*, 154:1, 2009.

Woods, Kevin M., 'The Iraqi perspective report – Saddam's Senior Leadership on Operation Iraqi Freedom', from the official U.S. Joint Forces Command report, Maryland: Naval Institute Press, 2006.

Woods, Kevin M., *The mother of all battles – Saddam Hussein's strategic plan for the Persian Gulf War*, Maryland: Naval Institute Press, 2008.

Woods, Kevin M., W. Murray and T. Holaday, *Saddam's War – An Iraqi military perspective of the Iran–Iraq war*, Washington DC: National Defense University Press, 2009.

Woodward, Bob, *Plan of Attack*, New York: Simon & Schuster, 2004.

Wright, Quincy, *A Study of War*, Chicago: University of Chicago, 1965.

Wrochem, Oliver von, *Erich Von Manstein: Vernichtungskrieg und Geschichtspolitik*, Krieg in der Geschichte, Paderborn: F. Schöningh, 2006.

WuDunn, Sheryl, 'China says Soviets Erred Earlier in Picking Leader' in *New York Times*, 8 September 1991.

Xiaoping, Deng, *Selected Works of Deng Xiaoping*, Vol. 1, 1938–1965, Beijing: Foreign Languages Press, 2006.

Zhenjiang, Fan, Zhao Tianliang and Jiefangjun Bao Zhang Guoyu, 'Military Theoretical Innovation Needed for Preparing for Information War, High-Tech War' in CPP20030121000089, *Beijing Jiefangjun Bao* (Internet version, in Chinese), 21 January 2003.

Zhou Fangyin, 'The Effect of the Information Revolution on Military Affairs and Security' in CPP20010817000186, Beijing Xiandai Guoji Guanxi (in Chinese), 1 August 2001.

Zhu Guangya interview featured in 'Watch Closely the Revolution of Military Technology in the New Era', in FTS19951023000001 Beijing Jiefangjun Bao, (in Chinese), 23 October 1995.

Zisser, Eyal, *Assad's Syria at a Crossroads*, (in Hebrew), Tel Aviv: Hakibbutz Hameuchad, 2007.

Index

Page numbers in **bold** denote figures.

Printed by Publishers' Graphics Kentucky